COMPUTER-AIDED SCULPTURE

COMPUTER-AIDED SCULPTURE

J. P. Duncan

Professor Emeritus
Mechanical Engineering
University of British Columbia

K. K. Law

Software Engineer
MacDonald Dettwiler
Vancouver, British Columbia

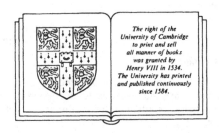

The right of the
University of Cambridge
to print and sell
all manner of books
was granted by
Henry VIII in 1534.
The University has printed
and published continuously
since 1584.

Cambridge University Press

Cambridge

New York   New Rochelle   Melbourne   Sydney

CAMBRIDGE
UNIVERSITY PRESS

University Printing House, Cambridge CB2 8BS, United Kingdom

Cambridge University Press is part of the University of Cambridge.

It furthers the University's mission by disseminating knowledge in the pursuit of
education, learning and research at the highest international levels of excellence.

www.cambridge.org
Information on this title: www.cambridge.org/9780521363037

© Cambridge University Press 1989

First published 1989

A catalogue record for this publication is available from the British Library

Library of Congress Cataloguing in Publication data

Duncan, J. P.
Computer-aided sculpture / J. P. Duncan, K. K. Law.
    p.   cm.
Bibliography: p.
ISBN 0–521–36303–9
1. Computer graphics – Computer programs. 2. Machine-tools –
Numerical control – Computer programs. 3. Polyhedral NC (Computer
program)   I. Law, K. K. II. Title.
T385.D86   1988
006.6'86 – dc19   88-39464
                                                                  CIP

ISBN 978-0-521-36303-7 Hardback

CONTENTS

PREFACE

This book presents in a concise form the basic theory of the POLYHEDRAL NC® system and computer programs for automatically defining and physically sculpturing surfaces of arbitrary shape. Documentation of these programs with demonstration examples is appended.

The underlying process may aptly be described as 'computer-aided pointing', the technique of classical sculpture. The system was conceived before high speed micro-computers with large memories became generally available. Early developments on mainframe computers were described by Duncan and Mair (1983) in *Sculptured Surfaces in Engineering and Medicine* and in publications listed there.

The storing, recalling and processing of data characterising this approach to surface replication can now be readily performed with practical advantages on mini- and micro-computers. This book describes, using IBM PC-compatible software, the updated versions of former mainframe programs with the addition of new ones for *volumetric property* computations to the former *surface* calculations.

Now data can easily and automatically be collected in fractions of a second by various laser-scanning measuring systems. Data so collected in a random format, or by Computed Axial Tomography Scanning (CAT Scan) now found in most hospitals, are reduced to a manageable format by the program TRUEPERS. The properties of surface-enclosed volumes and masses for arbitrary bodies, which are often required, are found readily by a new TETRAHEDRAL CONCEPT via the program VCAM (Volume, Centroid, Area, Moments) devised by the authors. A completely new non-interfering machining program, SUPERSUE, is introduced for the first time.

These new updated programs and demonstration examples are available on 5 $^1/_4$ inch IBM PC-compatible disks. All have been tested through the stages of definition to machining and their documentation is provided in the appendixes to this book.

April, 1989

J. P. Duncan
K. K. Law

ACKNOWLEDGEMENT

The development of the POLYHEDRAL NC® system was made possible by research grants from the National Research Council (NRC) of Canada and the National Science and Engineering Research Council (NSERC) of Canada over the years 1969-1987. The University of British Columbia and the University of Victoria, B.C. provided the computing and laboratory facilities. Initial development of the POLYHEDRAL NC® programs was assisted by Mr. John Hanson and Mrs. S. G. Mair, Programmer/Analysts. Many of the authors' colleagues and students, associates and technicians, too many to name individually, cooperated with the principal author from the outset. Their support and assistance are gratefully acknowledged.

Chapter 1

INTRODUCTION TO POLYHEDRAL CONCEPTS

1. The POLYHEDRAL NC® System

POLYHEDRAL NC® is a registered trademark in Canada for an automatic and universal machining method supported by a progression of computer-aided techniques and programs. The method involves both conceptual and detailed design of industrial or medical products, having either regular (analytical) or irregular and arbitrary shapes, and their formation by machining. It is thus suited to the sinking or formation of industrial dies and forms and to the replication of human anatomy as required in the manufacture of prostheses. Other applications are seabed and landscape models, floating hulls, archaeological artifacts and a number of unexpected items to be illustrated.

The system resides in some sixteen computer programs (executable on I. B. M. compatible mini or micro personal computers) on five $5\,^{1}/_{4}$ inch floppy disks with documentation. Output may be in the form of computer graphic presentations or as solid models formed in a computer-aided milling machine.

2. Polyhedral Machining

Polyhedral machining is a form of sculpturing akin to 'pointing' in classical artistic sculpture. A gouge-like tool 'digs' cavities close together on a block of raw material at the levels of the desired finished surface. Artistic sculptors use four-point divider devices to check the depth of the latest cavity against three previous ones. In POLYHEDRAL NC® both the depth and the position of a cavity are determined by pre-calculations in a computer. Thus this process could be described as 'computer-aided sculpture'.

A network, grid, or matrix of surface-point position-vectors is defined in the domain of the arbitrarily defined curved surface. The joining of these

1

points in suitable sets of three forms an inscribed/circumscribed irregular polyhedron approximating to the continuous surface. The triangular faces or facets of this polyhedron are touched in turn by the spherical cutting end of a ball-ended milling cutter as it is directed to traverse the whole field of the surface, removing material above the defined surface and leaving many cusps and grooves tangential to the facets. A close approximation to the surface is thus formed ready for 'finishing to witness' by hand-removal of the cusps or asperities. If the defined points are close enough, hand finishing may be unnecessary.

The other unique feature of the system, due to its theoretical basis, is its ability to avoid interference at points of high positive curvature. That means that while cutting correctly at one location in a surface-field, the tool will never remove material required for the correct formation of the surface at another location. Related to this capacity is the system's ability to fillet automatically at the slope-discontinuous junction of two or more surface-pieces, continuous in themselves and contiguous at their spatial curves of intersection. This is a compound-surface feature often met in conventional diemaking and pattern making which, in the latter craft, used to be dealt with by 'pattern maker's thumb' — running the thumb over a fillet of wax laid into the junction.

While suited to the formation of all types of single valued surfaces, the POLYHEDRAL NC® method suits particularly the geometrical characteristics of those surfaces that cannot be generated by the continuous motion of a spinning tool as in a lathe or on a shaping or milling machine. The generating process is always preferred where possible, for it gives a much smoother finished surface. Where generation is not possible, 'pointing' in some form becomes essential and POLYHEDRAL NC® provides a solution.

3. Elements of the Process

The process of POLYHEDRAL NC® machining involves three steps:

(i) The surface to be replicated must be defined as a large set of points in space, each lying in the surface. The points may be located from an origin by a Cartesian vector having components of length, x, depth, y, and height, z, measured from three mutually perpendicular planes forming a solid right angle. (Cylindrical coordinates may be used for closed or tubular surfaces as will be explained later.) Projections of such points in the xy plane may be in orthogonal, curvilinear, or random array as in Figure 1.1.

(ii) The surface specification in terms of points fixes the vertices and dimensions of a many-sided or many-faceted irregular polyhedron. By joining adjacent sets of three points, triangular plane-facets of the polyhedron are formed. Such facets may be thought of as the facets of a cut gemstone. Figure 1.2 shows a piece of surface confined to the first octant of space with some randomly distributed points marked on it. These have been joined to form a random array of contiguous plane triangles lying just within the curved surface. The ball-end of a milling cutter is shown just touching one facet tangentially. It is 'pointing' that facet at the instant shown. Figure 1.3 shows a surface-piece interpolated from the original random points to an organised array of points, each defined by Cartesian vectors and each having a height z above a base reference plane at location x, y. It may be thought of as representing a portion of a sphere having latitudes and longitudes as for the planet Earth. Here there are rows and columns in the topologically rectangular projection

Figure 1.1 Random and organised data.

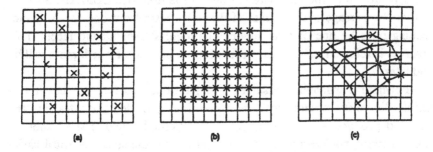

(a)   (b)   (c)

Figure 1.2 Random points of a surface joined to form a polyhedron of triangular facets.

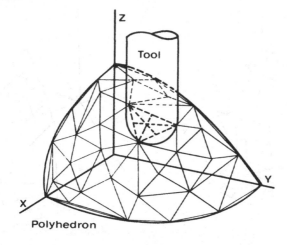

onto the *xy* coordinate plane. Thus the data *x, y, z* can be conveniently computer-stored at each node of the projected grid by reference to *x, y* as parameters or *u, v*, longitude and latitude.

(iii) The centre of the sphere of the tool must be placed at a point T so that it touches the facet 123 at its centroid C, a point in plane 123 chosen as both logical and convenient. It can easily be appreciated from Figure 1.4 how planes such as 123 can be defined from recalled data for points 1, 2, 3 and for all similar planes formed by joining diagonals such as diagonal 13 within each cell. A program, originally called SUMAIR but now NEWERSUE, pre-computes the positions such as T for each facet and ensures that while the tool touches plane 123 it will not interfere with any other plane within reach from T. If it does when the tool is placed at the normal position T, the tool is raised in direction *z* to a point T' so that the condition is avoided. The position for a smaller tool to be sent in later to get closer to the desired surface is computed and activated after the complete passage of the first tool. SUMAIR also organises the passage and progress of the tool centre from one touch to succeeding ones by one of four 'modes of progress' over all the facets. This progress is referred to as 'visiting the facets', a process of which more will be said later.

Polyhedral machining assumes a single-valued surface so that it can handle die cavity surfaces with the usual draft provided for extracting the moulded product. A special variant of the method called 'The Method of Non-Invasion', not presented here, is available for the strictly vertical cavity

Figure 1.3 Surface-pieces defined above topologically rectangular grids.

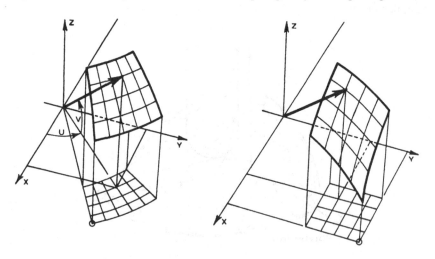

wall that is sometimes used for rubber-moulding dies, where the product can be removed by bending or distortion.

4. Properties of Polyhedral Surfaces and Their Enclosed Volumes

(i) *The surface area* of a polyhedron may be computed by summing the areas of all of its facets. This sum will be an approximation to the area of the continuous surface represented by the polyhedron provided that the facets and the sagittae of the facets are all small relative to the overall dimensions of the surface.

Each facet has vertices 1, 2, 3 with coordinates (x_i, y_i, z_i) where $i = 1, 2, 3$. The centroid of the facet has coordinates

$$x_c = (x_1 + x_2 + x_3) / 3$$
$$y_c = (y_1 + y_2 + y_3) / 3$$
$$z_c = (z_1 + z_2 + z_3) / 3$$

The facet 123 has an equation $ax + by + cz + d = 0$ and it is readily shown that $a, b, c, d,$ are given by determinants

$$a = \begin{vmatrix} y_1 & z_1 & 1 \\ y_2 & z_2 & 1 \\ y_3 & z_3 & 1 \end{vmatrix} \qquad -b = \begin{vmatrix} x_1 & z_1 & 1 \\ x_2 & z_2 & 1 \\ x_3 & z_3 & 1 \end{vmatrix}$$

Figure 1.4 Spherical tool touching a triangular facet.

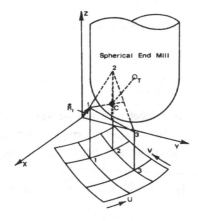

$$c = \begin{vmatrix} x_1 & y_1 & 1 \\ x_2 & y_2 & 1 \\ x_3 & y_3 & 1 \end{vmatrix} \qquad -d = \begin{vmatrix} x_1 & y_1 & z_1 \\ x_2 & y_2 & z_2 \\ x_3 & y_3 & z_3 \end{vmatrix}$$

The direction cosines of the normal to the facet (at the centroid as well as at all other points) are given by

$$\alpha = a / \sqrt{(a^2 + b^2 + c^2)}$$

$$\beta = b / \sqrt{(a^2 + b^2 + c^2)}$$

$$\gamma = c / \sqrt{(a^2 + b^2 + c^2)}$$

The modulus of the perpendicular distance of the facet 123 from the origin is given by the modulus of

$$p = d / \sqrt{(a^2 + b^2 + c^2)}$$

The outward (upward) normal, tending in the z positive direction, has a γ positive; the inward (downward) normal has γ negative of the numerical value of γ.

The area A_p of the projection of the triangular facet 123 on the xy coordinate plane at $z = 0$ is given by the determinant

$$A_p = 0.5 \times \begin{vmatrix} x_1 & y_1 & 1 \\ x_2 & y_2 & 1 \\ x_3 & y_3 & 1 \end{vmatrix}$$

Then the area, A, of the spatial facet 123, inclined to the coordinate axes by arc cos γ, is

$$A = A_p / \gamma$$

(ii) *The volume under a facet* of a polyhedron between the triangular plane of the facet and its projection on the xy plane may be found, as is conventional in the continuous mathematical process of double integration, by quadrature.

The volume is $V = A_p \times z_c$ where z_c is the coordinate of the centroid as stated above.

Then the total volume under a 'lateral' bounded surface as in Figure 1.4 can be found by summation of all such prismatic volumes.

(iii) *The volume enclosed by a closed (convex) polyhedron* is the sum of the volumes of all tetrahedra formed by joining all three vertices of a facet to a (suitable) internal point, P, as in Figure 1.5.

The volume, V, of elementary tetrahedra such as (123P) is given by determinant

$$V = (1/6) \times \begin{vmatrix} x_1 & y_1 & z_1 & 1 \\ x_2 & y_2 & z_2 & 1 \\ x_3 & y_3 & z_3 & 1 \\ x_P & y_P & z_P & 1 \end{vmatrix}$$

If the volume is fully enclosed by a complete set of contiguous triangles, the facets of a closed polyhedron, its value is the sum of all the elementary tetrahedral volumes as given by the above determinant. (The partially convex, partially concave polyhedron requires special consideration as discussed later.)

The other mechanical properties of the enclosed volume – centroid location, moments of volume, second moments of volume and products of volume (moments and products of inertia), directions of principal axes of volume (inertia) and the principal moments of volume (inertia) – may all be obtained by other summations and executions of algorithms relating to each of the many elementary tetrahedra formed as described above. These computations form the principal elements of a body of theory and programs which we have called

Figure 1.5 Elementary tetrahedron formed by joining facet vertices to the origin.

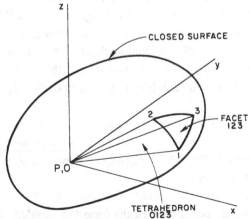

the TETRAHEDRAL CONCEPT. This concept is explained in detail in Chapter 6.

(iv) *The Cutter Location Data* (CLD) for a spherical tool caused to touch systematically, in turn, every surface-facet of a defined polyhedral surface, thus forming an approximate replica from a block of raw material, is obtained from

$$x_T = x_C + R \cdot \alpha$$
$$y_T = y_C + R \cdot \beta$$
$$z_T = z_C + R \cdot \gamma$$

where subscript C refers to the facet centroid, T to the centre location of the tool and R is the radius of the spherically-ended end-milling cutter used.

This CLD position, T, is a suitable position for the tool to 'point' the facet concerned under computer control.

Often the set of all positions T computed as above will cause the tool to touch each facet once only as shown for an extensive concave downward surface shown in two dimensional analogy in Figure 1.6. There is no interference.

If the surface has regions of positive (concave upwards) curvature of radius of curvature smaller than the radius of the spherical tool, interference may occur as illustrated in Figure 1.7. Then the tool is raised to a point T' so that either:
(a) the sphere is tangential to one high facet within the projected shank-circle of the tool but does not touch lower ones within possible reach (Program NEWERSUE). Or:
(b) the sphere is raised until all nodes (vertices of facets) within its possible reach at its current vertical centre line location are at a distance from the spherical centre greater than its radius (Program SUPERSUE).

The possibility of interference as a tool touches one facet is easily appreciated from the two dimensional analogue diagram shown in Figure 1.7. The shaded region would be removed before the tool reached the facets within it unless the tool is withdrawn to centre position T_i on its current axis location.

In case (a), the CLD for a 'cascade' of tools of ever decreasing radius (usually each being $^1/_2$ of the last used) is found automatically so that smaller and smaller tools may be 'sent in' to penetrate local regions of high curvature to form its facets correctly.

Some other systems avoid interference (for instance AUTO-GRAPHICS) but do not automatically form the avoided element of surface subsequently as does POLYHEDRAL NC®. In the latter system a very large tool may be used for 'hogging out' unrequired bulk material, knowing that nothing required later will be cut away. Tools of smaller radius used later will only be 'sent in' to those regions of the surface needing their attention.

(v) *The errors of POLYHEDRAL NC®*. The POLYHEDRAL NC® method is an approximate way of representing and machining a surface by reducing computations to linear analysis. Other advantages attend the method. Discontinuities are involved but so are they in the treatment of surfaces having known non-linear functional form: ultimately the processing and

Figure 1.6 Spherical tool touching a concave downwards curve in machining.

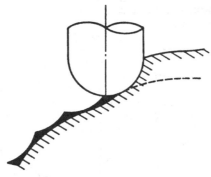

Figure 1.7 Spherical tool interfering in a 'groove' during machining.

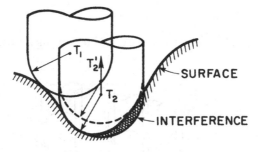

2-D ANALOGY OF RETRACTION
TO AVOID
INTERFERENCE

machining of these must be done in digital (non-continuous) computers which work in steps and increments; 1 and 0; yes and no.

The errors in POLYHEDRAL NC® are (a) representational and (b) errors of replication. The representational error is characterised by the three dimensional sagittae as indicated by a two dimensional analogy in Figure 1.8.

This displacement error is sometimes positive and sometimes negative according to the local curvature and its sign. It may be made numerically smaller and smaller by taking facet vertices ever closer together.

The errors of replication have to do with the disparity between the tool sphere and the facet plane and with the measures taken to avoid interference.

Figure 1.8 Spherical tool touching internal and external facets of a polyhedron approximating to a surface.

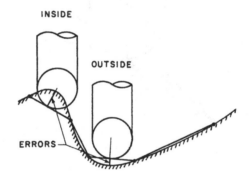

Figure 1.9 Cusps left by a spherical tool traversing a surface cross-section at intervals.

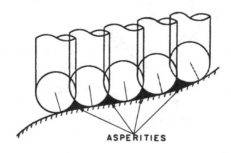

In Figure 1.9 showing a two dimensional analogue of what is really a three dimensional situation, a spherical tool is addressed in turn to a series of facets. Cusps or 'mountain peaks' are left above the desired surface after the tool has visited all facets.

Usually the nodal spacing of facet vertices will be about $1/2$ to $1/3$ of the tool radius or even less. The height of the cut surface after all tools have visited all facets needing their attention at each nodal location in x and y is easily calculated and a statistical assessment of these errors may be made. This has been done for and is incorporated in the output of program NEWERSUE.

A similar treatment has been applied to the alternative machining program SUPERSUE. In this approach, tools are always positioned with their axes passing through a node. The tool is raised until it just avoids cutting beyond any nodal point within reach when its axis is located at a particular node. Again the residual error at each node after all nodes have been visited may be computed and statistically assessed.

These errors can be controlled to a large extent by appropriate choice of facet size (grid spacing) and tool-sphere radius. The current availability of large Random Access Memory (RAM), and of modern measuring techniques capable of collecting data from physical surfaces at 1.5mm spacing over, say, a 20 x 20 x 20cm volume, makes the choice of small nodal and grid spacing quite feasible.

(vi) *Approximate Curvature* of a surface digitally defined above a parametric or topologically rectangular grid of rows and columns, (to which form even random surface data can be reduced by the program TRUEPERS) can be found by finite difference approximations. The independent program CURVATUR computes the maximum principal curvature at each and every node in a field and draws 'hills of curvature'.

The smallest radius of principal curvature may be found as the reciprocal of the maximum curvature. If it is desired to machine the whole surface without encountering anti-interference retraction, the cutting tool radius should be smaller than this smallest radius of curvature. If a tool radius has been arbitrarily chosen larger than this, the region of interference will be apparent from the curvature plot and the error routine will compute and reveal the errors which will result from the use of any specific 'cascade' of tools.

(vii) *Modes of progress*. When a surface has been defined by point position vectors in a topologically rectangular array (it need not be geometrically rectangular, see Chapter 2), one facet might look as in Figure 1.10. Point $P(x_1, y_1, z_1)$ is a typical node.

By joining P to Q we create two triangular facets QPS and QPR. By doing this in the same sense for all 'cells' between two rows and two columns we create plane facets bounding the whole surface.

We may choose to join for all cells *either* typical diagonal PQ *or* typical diagonal SR (as shown for SR in another cell).

Then taking a true plan view of the *xy* plane, viewing along the *z* axis in the negative direction we see cells and facets as in Figure 1.11.

Two quite different polyhedra with triangular facets are formed by joining the PQ set, Figure 1.11(b) and (c) or the SR set, Figure 1.11(a).

For Figures (b) and (c) two possible paths for tool visitation of facets are shown. These we call 'modes of progress'. In (b) the tool moves in zig zag fashion developing surface ripple-errors –'peaks'. In (c) the tool passes twice between each pair of rows in track-like progress, leaving a series of grooves as with conventional machining procedures.

Zig zag or tracking progress is, of course, possible with diagonals SR joined in each cell, giving in all four possible 'modes of progress' in facet visitation. These alternative modes can be selected within NEWERSUE.

In SUPERSUE tools are always positioned with vertical axes passing through a node. Then progress is from node to node along a row. Then a shift to the next row and another traverse of that row.

SUPERSUE thus obviates the necessity of computing increments of *x* and *y* (as is necessary with NEWERSUE). The *x* and *y* increments are always

Figure 1.10 Formation of facets above a topologically rectangular grid.

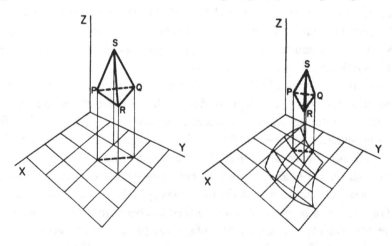

the same. The anti-interference measure used here is avoidance of surface-points rather than of facet-planes. There is possibility of undercutting the facet-planes adjacent to the projected circle of the tool-shank as will be apparent from Figure 1.12.

5. Conclusion

These are the main quantitative and qualitative features of the POLYHEDRAL NC® system. In ensuing chapters we trace in more detail the derivation of point-position vectors for surface representation and the theory of tool positioning for machining.

Figure 1.11 Modes of machining progress.

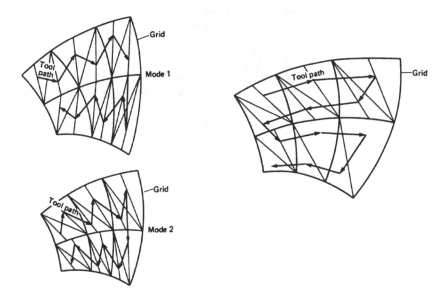

Figure 1.12 Avoidance of node points in machining by SUPERSUE.

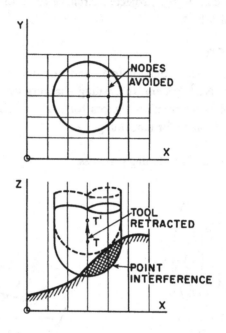

Chapter 2

DESIGNED SURFACES AND THEIR THEORETICAL BASES

1. Introduction

Surface is a theoretical concept which may have physical reality. Physical objects have a surface bounding a volume. Man-made objects have to be designed and that involves the determination of a bounding surface-shape. That shape will have an equation – a functional relationship between its Cartesian coordinates *(x, y, z)* or its parameters *(u, v)* – for surface is a two-dimensional *(u, v)* entity set in three dimensional space *(x, y, z)*.

A designed object is under the control of the designer. He or she chooses the function describing it. The function for a pre-existing object to be replicated must be found by measurement and derivations from the measurements.

We consider the designed object first and discuss the treatment of natural, physical objects in Chapter 3.

2. Surface Equations

There are three basic types of surface equation applicable to ordinary physical objects:

(i)   Classical (implicit) $F(x, y, z) = 0$
(ii) Mongean (explicit) $z = f(x, y)$
(iii) Gaussian (parametric) $x = f_1(u, v); \; y = f_2(u, v); \; z = f_3(u, v)$

POLYHEDRAL NC® makes of all three types but the general polyhedral theory is built around the Mongean form.

Gaussian parametric surfaces are used mostly by applied mathematicians, since they are axes independent, unambiguous and amenable to the methods of differential geometry. But surfaces so defined have, ultimately, to

15

be described in terms of Cartesian coordinates to suit the three axes and basic movements of machine tools. POLYHEDRAL NC® accommodates derivation and description by Gaussian parameters as well as Mongean description. This is because the facets of approximating polyhedra can be defined by any topologically rectangular matrix or grid. The rows and columns may be either $x = $ constant and $y = $ constant or $u = $ constant and $v = $ constant.

Figure 2.1 shows the natures of a Mongean and Gaussian surface-patch. In the case of the Gaussian description, z is still a function of x and y but both x and y, as well as z, are functions of u and v.

In computing and storing (x, y, z) coordinates of points in either surface description, the coordinates x and y may be incremented in scanning a field by constant increments (they are, essentially, the parameters) or u and v may be incremented by fractions of their ranges, usually unity. These parameters are not identified with the Cartesian coordinates. They are independent of them but determine their values.

The resulting projections on the xy plane of surface points at nodes of the two types of surface is as shown in Figure 2.1. We call these projections 'topologically rectangular'. In each case there are the same number of nodes in each row and in each column whatever the geometrical shape of those rows and columns may be. This feature facilitates computer storage and recall for processing.

Figure 2.1 Surface-patches defined by Monge's and Gauss' equation.

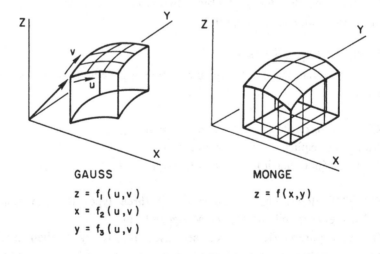

GAUSS
$$z = f_1(u,v)$$
$$x = f_2(u,v)$$
$$y = f_3(u,v)$$

MONGE
$$z = f(x,y)$$

The normal to a surface, which is needed for the positioning of cutting tools, is found conveniently from the classical form of equation as having direction cosines α, β, γ :

These relationships can be applied to any analytical surface defined by $F(x, y, z) = 0$.

Hence for a plane triangular facet, whose general equation is

$$ax + by + cz + d = 0,$$

the values of α, β and γ are as stated in Chapter 1. These were derived from the general classical equation applied to the plane.

3. Types of Surface Description

It is not intended to expound surface mathematics and theories here but rather to state and explain some approaches to and types of surface that have been implemented via POLYHEDRAL NC®.

These types include:

(i) Surfaces with implicit or explicit functional form in Cartesian coordinates.

(ii) Surfaces described in Gaussian parametric form whose shape is constrained by various stated point-positions and slopes such as Bézier bi-cubic patches.

(iii) Surfaces required to 'bridge' between twisted curved spatial boundaries treated by the method of Proportional Development.

(iv) Surfaces generated by superposition of local functions – bi-beta and B-spline type – which are also used for surface adjustment.

In Chapter 3 the interpolation of measured random data from contouring and sectioning of natural, physical surfaces and the fitting of a polyhedron to these data will be specially treated.

4. The Method of Highest Point

Many die-cavities and punches are defined geometrically as a set of analytical surface-elements intersecting each other at junctions – twisted space-curves – which are filleted to provide localised continuity. The whole assembly of elements is thus a completely slope-continuous, compound, analytical surface.

The Method of Highest Point is a way of deriving vectors $\mathbf{r} = x\,\mathbf{i} + y\,\mathbf{j} + z\,\mathbf{k}$ which define the uppermost point at a specified location in plan (x, y) of points of a set of surfaces defined independently in a given domain. If these independent elements are analytical, the test of highest point can be made easily by evaluating z for all surface-elements which may exist at a given location *(x, y)* in the plan projection (using Monge's equation) and determining the highest value. (In some cases the lowest value may be appropriate.)

By conducting this test systematically at each location (x, y) – the nodes – of a topologically rectangular grid in the $x\,y$ coordinate plane, the spatial nodes of a polyhedron consisting of uppermost points at each location may be found.

If any of the several independent surface-elements is defined by random point vectors or by a curvilinear but topologically rectangular grid (such as a Bézier surface), it may easily be reduced to definition in terms of a geometrically rectangular grid by program TRUEPERS to be described later. Then any such surface element can be tested against all others for the highest point by scanning the field by increments Δx and Δy.

The compounding of some well known surface-types is carried out automatically by BASIC computer program GEN7 which provides for the assembly of up to 37 elements. These embrace all the surface-types usually discussed in works on calculus and analytical geometry – planes, ellipsoids, paraboloids, hyperboloids and their special cases, cones, cylinders, etc., – to which has been added the torus.

The details of the Method of Highest Point are given in Chapter 9 of *Sculptured Surfaces in Engineering and Medicine* (Duncan and Mair, 1983), which deals with the general topic of Compound Surfaces.

Figure 2.2 shows the compounding of a plane, two cones and two tilted circular paraboloids to define the form of an automobile tail-lamp reflector punch. The array of highest points is presented in true perspective as a topologically and geometrically rectangular array of surface-points, each defined by a vector from an origin and each having a z value at each plan-view point (x, y).

Figure 2.3 shows the arbitrary compounding of an inclined plane, a circular cylinder, a sphere, a paraboloid and a cone. As in all illustrations of the Method of Highest Point, hidden lines have been removed by program PLOT3D (plotting section of TRUEPERS). The linkings of computed surface-points form two sets of orthogonal profiles. Contours may be obtained by call-up if required.

The documentation of computer program GEN7 explains the method further.

5. The Bézier Patch and Surface Quilts

There are a number of methods based on parametric surface vector-equations and methods for rendering partially defined surfaces amenable to analysis. These methods are associated with the names of Coons, Bézier, Ferguson, Gregory, and others. We will use a Bézier's bi-cubic patch to explain how definitions of surfaces in terms of parameters and vectors can be linked with POLYHEDRAL NC®.

Each of these methods calls for an initial vector specification of some relatively widely spaced points lying in the surface through which sets of intersecting surface-curves are supposed to pass to form patches of surface-elements as shown in Figure 2.4 . The corner points are called (special) nodes;

Figure 2.2 Automobile tail lamp punch. The principal sections of an automobile tail lamp punch were extracted from a commercial drawing. The polyhedron including two paraboloids, two cones and one inclined plane emerged by application of the Method of Highest Point.

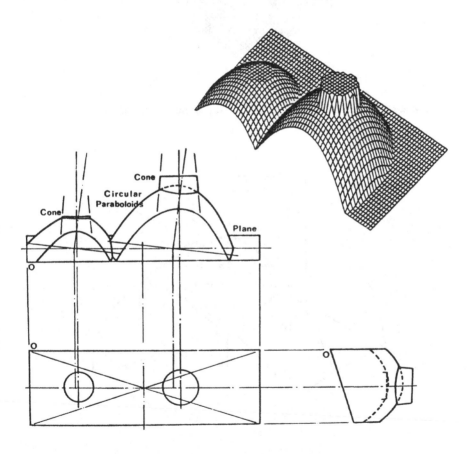

Figure 2.3 Arbitrary compound of analytical surface.

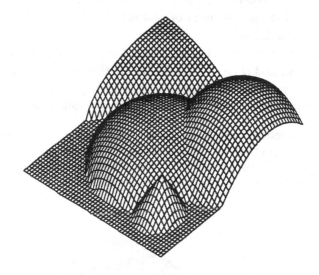

Figure 2.4 Surface composed of analytical patches – a 'quilt'.

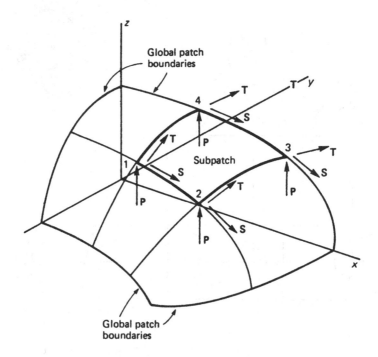

the twisted space-curves passing between them, patch boundaries. Other curves within the patch form a 'net' of twisted space curves lying in a bridging' surface and defining it. Their intersections comprise other (general) nodes. From the figure it can be seen that the projections of each such node onto the xy plane form there a topologically rectangular grid or array, the linkings of which are, however, curvilinear. So this array is not *geometrically* rectangular.

In the various methods the patch is represented by bi-parametric expressions, the parameters commonly being u, v ranging from 0 to 1 along adjacent bounding curves passing through a reference corner node.

A point in a surface-patch is defined by a vector $r(u, v)$ and the surface itself is continuous, usually up to the second degree except at boundaries of patches. In the various methods, particular functions $r(u, v)$ are specified by vector equations. Boundary conditions are usually given at corner nodes – displacements and slopes (perhaps curvatures) – and, according to the degree of the equation chosen, displacements at a limited number of surface points within the boundary are arbitrarily specified.

The surface shape can only satisfy a limited number of conditions within the patch but numerous patches may be assembled with position and slope continuity at nodes as illustrated in the 'quilt' of patches shown in Figure 2.4 . Then both the corner nodes and internal nodes of each patch all have given or computed z values at x, y positions in a topologically rectangular grid projected in the $x\,y$ plane. Such patches can then be machined according to POLYHEDRAL NC® procedures using linear approximations associated with facets rather than vector products of parametric slopes to find the normals to the surface.

A typical Bézier patch is illustrated in Figure 2.5 in which the maximum of four independent internal nodes of the surface are shown. There are 16 coefficients Q_i to be found. Frequently four corner nodal locations are given together with two partial slopes at each corner, giving a total of eight conditions. Thus four more positions at most can be arbitrarily given for surface points within the boundary.

The above treatment does not ensure curvature continuity. If higher degree vector equations are used this might be possible. Alternatively, a limited number of curvature conditions might be substituted for point positions.

The theory of Bézier and other types of vector surface equation patches has been thoroughly developed elsewhere. There is no need to pursue it further here. The significant point to note is that vector-defined surfaces of this type can be viewed and treated as polyhedra with nodes as vertices. Normals can thus be found approximately from analysis of facets and, as will be shown, approximate curvatures can be found for assessing machining feasibility by finite difference calculus.

Another important point is that the program TRUEPERS can search a computer memory for the Cartesian components of the u, v vectors of surface-points and, *treating them as random*, transform the curvilinear, topologically rectangular grid of projected points into a geometrically rectangular set as in Figure 1.1. Then a Bézier vector patch could be integrated with other surfaces, using that common array, by the Method of Highest Point.

6. Proportional Development: PROPDEV and CURVFIT

In practical die design, the elements of a surface are often bounded by a closed spatial curve, the designer or toolmaker (in past practice) being left to 'blend' a surface within this boundary. Frequently, as is still the case, this closed spatial boundary is comprised of several elements of plane curves in the principal planes of a Cartesian space. Some of these curves may lie in planes of symmetry of a surface defined in two or more octants of space: some may constitute 'raw edges' of a panel where slope-matching with adjacent surfaces if any is not required. The surfaces may only have to be contiguous.

Figure 2.5 The 'traces' (circumscribing polyhedron) of a Bézier patch.

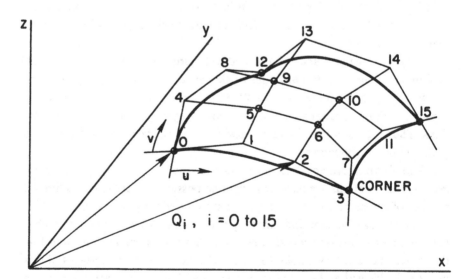

An example of such a configuration is the automotive roof panel shown in Figure 2.6 .

Yet other surfaces may have four twisted space curves joined at four corners. The four curves do not have to be discontinuous at these corners: they may form a smooth, continuous, closed curve on which four points may be 'marked' for recognition as corners (to render the example similar to a conventional vector-defined patch).

In Figure 2.7 we show a four-cornered patch projected onto $x\,y$ and $x\,z$ coordinate planes. These projections are then 'folded' about the x axis to form two plane projections of the boundary.

In Figure 2.8 the patch is assumed to have a continuous spatial boundary. Four 'corners' have been arbitrarily selected and the projecting and folding process repeated as above.

These figures show that a closed space-curve can be represented by two plane projections. (Ambiguities can arise but will be obvious in practical design – Duncan and Mair (1983), Appendix C.)

Now by setting the axes of y and z vertically upwards to suit conventional graphical plotting we have a boundary representation in two plane, closed curve projections as in Figure 2.9 .

Figure 2.6 Rear automobile roof panel defined as a 'quilt' of contiguous and continuous patches.

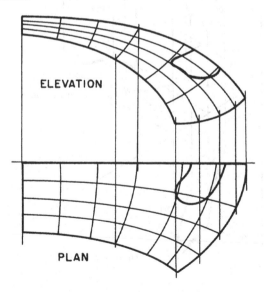

Also in this figure we have geometrical constructions for finding the coordinates of the position in each view of any and all points within the closed real boundary which constitute a continuous smooth surface whose smoothness is related to that of the four boundary curves.

The parameters of this surface are selected to be fractions of a range of coordinates between the elements of the boundary passing between the corners. There are two parameters possible – α and β – (sometimes, fortuitously, three) and these can be applied in four alternative ways.

First, each projected boundary element (there are eight of them) must be functionalised – expressed by an equation. Sometimes this may be done by giving the element an arbitrary function, such as the equation for a straight

Figure 2.7 Surface-patch boundary projected onto Cartesian planes.

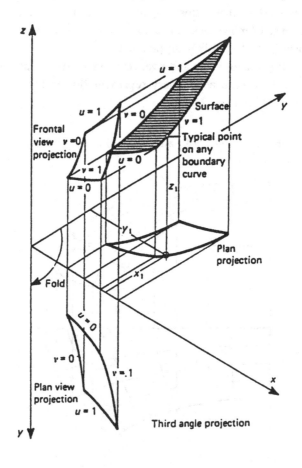

line or parabola. In conventional orthographic methods of design, which are still suited to a person's ability to think about and plan three dimensional objects, a boundary curve may be freehand-sketched. Then it may be digitised on a modern flat-bed digitising tablet and curve-fitted by any of many computer-aided methods.

The authors use a concept and program of their own devising called CURVFIT. This receives digitised data for a plane curve, the coordinates of an inscribed/circumscribed polygon, and fits to these data, automatically, a train of tangential general second degree arcs. The fit can range from 'faithful following', by passing through every point, to 'smoothing', by specifying a permitted root mean square departure from the polygon. Second degree arcs are

Figure 2.8 A continuous surface-patch boundary projected onto Cartesian planes.

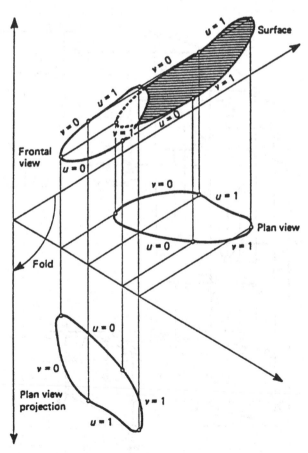

used because they cannot oscillate or give 'ripples'. Inflexion points are always selected and controlled as are maxima in both x and (y or z). Multi-values in the curve-fitting elements are taken care of by various devices as are vertical tangents (slope = infinity). The pieces of arc are joined with tangential continuity, the automatic shortening of them to suit the specified degree of fit being controlled by another device – weighted mean slope.

The documentation for CURVFIT explains the above in more detail and examples of the application of CURVFIT permeate this work.

CURVFIT is the boundary-defining method used by the authors in the process of Proportional Development of surfaces. This method is fully explained in *Sculptured Surfaces in Engineering and Medicine* (Duncan and

Figure 2.9 General scheme of proportional development of a surface-patch with boundaries given as Cartesian projections.

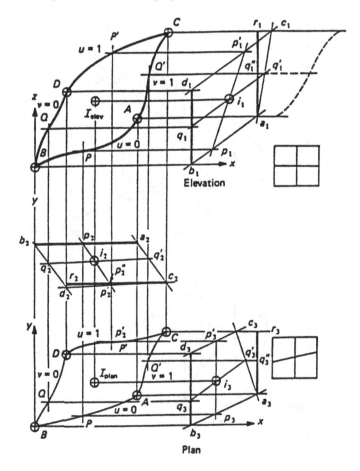

Mair, 1983) with many graphical examples. In past days of automotive body design, this whole procedure was carried out graphically, but today it is executed by computers through algorithms.

When completed, the array of point-position vectors defining a surface as finely and continuously as desired from the Method of Proportional Development resembles or may resemble a Bézier array in Cartesian components except that the parameters are α and β, fractions of two of the coordinate ranges.

The bridging surface spanning the closed boundary has one of four set shapes fixed by boundary specifications. If the boundary elements are continuous, so the surface will be (to the same degree of continuity). If now a smooth adjustment of this shape is desired, we apply by superposition a bi-beta adjusting surface as introduced below. In Bézier surface development internal points (four available) are adjusted and the development is iterated.

Thus the Proportionally Developed PROPDEV surface, with adjustment if necessary, leads to an array in which $z = f(x, y)$ to which form the adjusted Bézier surface may also be reduced for processing by the POLYHEDRAL NC® system, Program TRUEPERS (see the demonstration programs herewith).

7. Bi-beta Functions and B-splines

The bi-beta and B-spline types of curve and surface provide yet another approach to surface design but also a means of adjusting a developed surface to suit special requirements or criteria. The general principle of these curves and surfaces is that a function contributes to a limited range or region of a curve or surface about a centre – a 'knot'. The whole curve or surface is then defined by superposition of many such elementary curves or surfaces covering the whole interval or surface-patch. In the case of adjustment, the superposition of an adjusting element may be applied either to the whole or to a localised region of the curve or surface.

The beta function for a plane curve has the well known equation for a beta distribution in statistics:

$$z = A_n x^{\lambda 1} (1-x)^{\lambda 2}$$

where x ranges from 0 to 1. For a central maximum ordinate, $\lambda_1 = \lambda_2$. For a non-central maximum, $\lambda_1 \neq \lambda_2$. If $\lambda = \lambda_1 = \lambda_2 > 2$ boundary slopes and curvatures at both ends of the curve are both zero.

Figure 2.10 shows the forms of the beta function for various values of the powers, λ. Clearly the curves have non-zero ordinates only within the interval 0 to 1.

An extensive curve may be 'built-up' from many beta functions centred at a series of knots as shown in a simple way in Figure 2.11. The ordinates of the extensive curve are made equal to the sum at any given location of those of the beta functions which have non-zero ordinates at that location. Thus each

Figure 2.10 Shapes of beta functions for various combinations of λ. (Courtesy of K. Bury.)

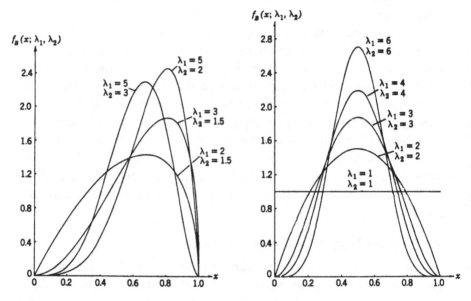

Figure 2.11 Superposition of beta functions.

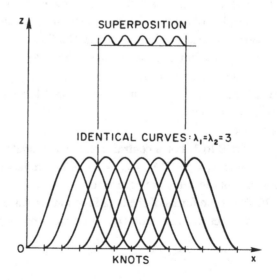

elementary beta function has a limited range. The coefficients A_i, where $i =$ 1, 2, 3, . . ., n, of the elementary functions are found by solving a set of linear equations set up at each 'knot'. The ordinates of the extensive curve can then be found at all points by recursive formulae. Some form of interpolation between the knots will result.

This very elementary example of curve fitting indicates that beta functions (and B-splines, which employ a similar, more general approach) are, effectively, *influence functions*, distributed along the extensive curve but influencing only a limited interval along the curve.

Here we have indicated only the attainment of *displacement* along the curve. Nothing has yet been said about slopes and curvatures either at the knots or between them.

By creating a beta function in two coordinate directions simultaneously, we obtain a bi-beta function as follows:

$$z = A_n x^{\lambda 1}(1\text{-}x)^{\lambda 2} \ y^{\lambda 3} \ (1\text{-}y)^{\lambda 4}$$

where $\lambda_1, \lambda_2, \lambda_3, \lambda_4$, may all differ from each other. This function gives a surface within a unit square as in the inset of Figure 2.12. The maximum height is

Figure 2.12 Bi-beta function 'spread' into the outline shape of a violin.

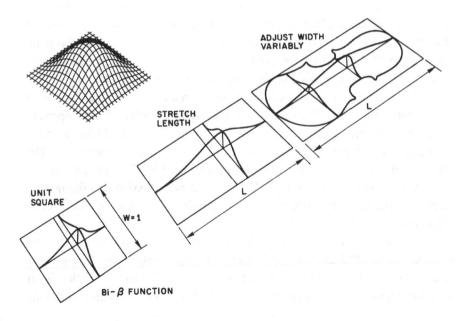

ADJUST WIDTH
VARIABLY

STRETCH
LENGTH

L

L

UNIT
SQUARE

W = 1

Bi-β FUNCTION

always at some location $x = x_1, y = y_1$. For $\lambda > 2$, boundary displacements, slopes and curvatures are zero everywhere. The bi-beta function has 'influence' only within the unit square or in any local region of an extensive field to which it may be applied by scaling.

The bi-beta function represents a 'pimple' or 'dimple'; many of which could be applied at a series of surface 'knots' for superposition to form an extensive surface above an existing plane or curved surface to adjust its shape to suit some requirement.

Suppose we are given a Bézier (or proportionally developed) surface-patch whose parameters u, v (or α, β) range from 0 to 1 as sketched in Figure 2.5. At each point u, v there will be an ordinate $z = f_3(u, v)$. These could be plotted in a three dimensional graph. This plots the ordinates z above a unit square. Thus a bi-beta function could be directly superimposed on the Bézier patch with maximum ordinate at any chosen location (u_1, v_1) to influence the whole surface or, by scaling the bi-beta function, to influence a limited region determined by chosen ranges of parameters u, v. Thus the original surface may be adjusted until, after inspecting screen plots of the results of superposition, a modified surface closely correlated to special criteria may be defined. Figure 2.13 shows these processes graphically.

A detailed example of the use of bi-beta surfaces to adjust the mould lines of a ship's hull is explained fully in Duncan and Mair (1983).

For a second example we describe the use of a bi-beta function for designing (and machining) an experimental top-plate for a violin as shown in Figure 2.12. In this we use one bi-beta function 'spread' into the special plan-view shape of that instrument (and another to adjust it locally as required near the bridge) as explained in Duncan, Lau, and Steeves (1984).

We take for a boundary of the plate the outline of a good instrument traced out on a plane. The doubly curved extensive surface of the top-plate (bottom-plates are similar) has to have certain 'arching' and thickness variation and is traditionally hand-carved out of solid planks of pine wood. The wood used – its grain and quality – influences the desirable arching; carving takes several weeks of hand-work. If a series of plates having definite mathematical specifications could be made quickly by machining, instrument development would be assisted.

The outline of the plate is first digitised and functionalised by the program CURVFIT as shown in Figure 2.14. A parallel, inside margin-curve is also functionalised to bypass the C bouts. In these fittings, correlation is set at a fine root mean square (RMS) limit and CURVFIT determines and finds the number and coefficients of the necessary pieces. So the boundary of the plate is piecewise functionalised.

Next, a bi-beta function with its maximum ordinate on the centre line of the plate but slightly closer to the finger board end than to the bridge end, as required by tradition, is postulated and its λ's are given appropriate values, guided broadly by traditional configurations.

The unit square in which this function exists is then 'stretched' in the direction of its axis of symmetry to span the length of the top-plate as shown in Figure 2.13. Then using the boundary function already found, the bi-beta function is stretched laterally but variably along the length of the top-plate until the original square maps exactly into the plan-view shape of the violin. When points determined by the bi-beta function are thus 'moved' to new plan-view positions, they take their z values with them. Thus we generate a bi-parametric surface, the parameters being those between 0 and 1 in the

Figure 2.13 Superposition of bi-beta surface on a bi-parametric surface (for adjustment).

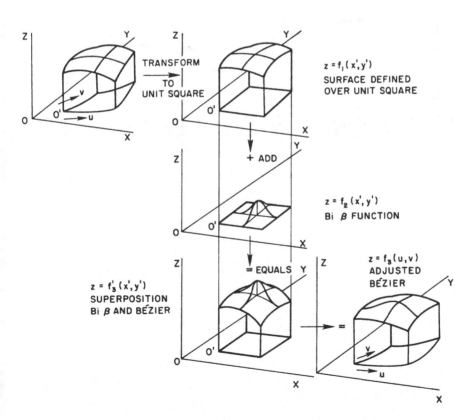

original unit square. The developed surface will be as smooth as the original
bi-beta function and is defined continuously everywhere. Effectively the
original function has been scaled, uniformly in one direction and variably in
the other.

A good violin has to have a certain thickness gradation as well as the
asymmetrical thin spot near the bridge. The thickness can be controlled by
another, similarly 'spread' bi-beta function defining the inside surface.

B-spline curves and surfaces are developed by treatments analogous to
but more refined and detailed than the treatments given above. A *mixture* of
functions of limited influence as in Figure 2.11 is used. The method is now
much favoured as a general approach to curve- and surface-fitting and could
now be called a standard method, well documented in both elementary and
advanced literature.

Figure 2.14 Violin top-plate outline defined as a series of CURVFIT curves.

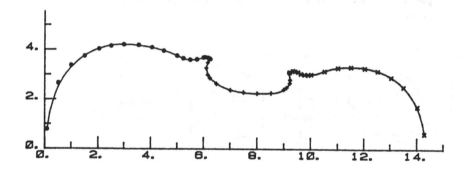

Chapter 3

NATURAL SURFACES AND DATA COLLECTION

1. Mechanical Measurement and Sounding

Natural surfaces to be replicated by methods of Computer-Aided Design and Computer-Aided Manufacturing (CAD CAM) for various utilitarian purposes have to be measured in some way at some discrete points and then interpolated to complete an adequate definition.

We think here of copying the artist's clay model, of modelling landscapes and seabeds, of shaping aesthetic dies and moulds or of forming human prostheses for the handicapped.

An obvious way of finding the coordinates of some random points on an arbitrary physical surface is to place the object on the table of a measuring machine and use a mechanical probe to touch the surface and record the coordinates. This assumes that the object is small enough to be placed on such a table. A hand operated measuring machine for this purpose can record coordinates to ± 0.00025mm over a field of 1000mm x 1000mm by compensating for the offsetting effect of the finite probe-tip diameter.

This last problem may be solved by fitting five local probe-ball centre-positions near a central point of interest with a right paraboloid of the form $z = ax^2 + by^2 + cx + dy + e$. Then by differentiating the resulting function, the tangential plane and normal may be found. Hence compensation for the offsetting of the probe-ball from the surface can be applied to find the coordinates of a surface point. Here the method of computing the cutting-tool offset for a spherical tool from a given surface-facet, as explained in Chapter 2, is applied in reverse.

For greatest accuracy in measuring true surface-points, this correction procedure should be applied. A surface can also be replicated by causing a spherical cutting-tool to visit the same coordinated points as are first recorded by a spherical probe of the same size as the tool during a virtually continuous scan of the prototype surface. This was the method used by Keller copying machines in the 1940's.

In another form of such mechanical measurement called 'the Reflex Plotter', the *virtual optical image* of delicate soft objects such as hearts, kidneys, etc. may be digitised. In this a point-probe of virtually zero tip-radius is caused to touch the *image* of the object rather than the object itself.

In another three dimensional digitiser known to the authors, the x, y, z position of a probe or pointer placed on a physical object is sensed electromagnetically by three receiver-stations which employ deduction of point coordinates by triangulation and techniques from modern land surveying methods.

Marine charts bear depth figures at a number of random points in the x, y field and these are usually obtained by acoustic sounding. Acoustic sounding is another technique which is finding more and more applications in depth and distance measurement. We show below how random coordinated points found by sounding, or alternatively by mechanical measurements, can be interpolated to produce a topologically rectangular array which in turn can be processed by POLYHEDRAL NC®.

2. Reflection and Scattering of Light at Surfaces

Since most modern measuring systems are essentially optical we will dwell on these.

Physical surfaces are seen in terms of the incident light which they either reflect or scatter to the eye of the observer from points on the surface. The microscopic surface-roughness and the constituent nature of the substance underlying the surface determine this response to incident light. Sir Isaac Newton made his telescopes of a metal called speculum which he highly polished. The regular, controlled reflection which he thus obtained was called 'specular', a term still used. A non-specular surface will scatter incident light in many directions from a point. If the surface is like human flesh, some incident light may be scattered from within the substance underlying the surface. This gives rise to some important differences between optical effects obtained with scatter from bounding surfaces of translucent and opaque substances.

Specular reflection and refraction are the basic mechanisms of classical interferometry; surfaces of relatively high curvature, such as are encountered in the present context, are usually measured by techniques dependent upon scatter. Holography is based on the scatter of coherent light but its execution is demanding and its sensitivity is often too great to be useful for routine application to engineering and medical surface-measurement. So we concentrate here on the use of a few non-coherent and readily manageable optical methods for collecting surface-point coordinate data with a view to replicating physical surfaces by machining.

3. Stereo-photogrammetry

This is one of the oldest methods of recording the topography of both large areas of terrain photographed from aircraft or of buildings and small objects viewed at close range, meaning focused at a finite distance less than infinity. Two photographs of the terrain or object are taken by cameras at nearby positions. The depth z of points in the scene at positions x, y in the plane photographs can be deduced from the photographs by measuring the parallax of identifiable points in them. Figure 3.1 indicates graphically the meaning of parallax.

Figure 3.2 shows some contours deduced from photographic stereo pairs. In photogrammetry, photographs are 'read' by a skilled, specially trained operator of one of many very expensive analysing machines which currently are tending towards ever more automation. The output may be z values at orthogonal rows and columns in x and y (plan view coordinates) recorded on computer-compatible storage media. Alternatively, the coordinates of random points may be recorded and interpolated subsequently to an orthogonal array in a plan view by program TRUEPERS. Also, the operator may 'read' the stereoscopic image continuously at a present level and thus cause a plotting accessory of the machine to draw contour maps on paper of the type shown in Figure 3.2 .

The art of stereo-photogrammetry is highly specialised and the surface replicator usually looks to such specialists and their very expensive machines to have photographs processed. Some researchers do it themselves on so called low order stereoscopic devices.

In this art of depth measurement by parallax-sensing, Charge Coupled Diode (CCD) chips and other forms of light-sensing arrays are now replacing conventional photographs as recording media. These enable point coordinate data to be fed directly to computer storage media. This is described below as a new category of recording methods based on laser scanning.

4. Shadow-moiré Contourography

This is a method that deduces depths of a surface from one rather than two photographs. Essentially it senses parallax and computes and plots contours of equal elevation as fringes superimposed on the single photograph. The method is about one order less accurate than conventional photogrammetry but is often more than sufficiently accurate for engineering or medical surface reproduction.

Duncan and Mair (1983) show many examples and give detailed descriptions of the shadow-moiré method. In its development by Meadows, Takasaki, Terada and others in the 1970s, the method is carried out on the basis of central perspective illumination and photography rather than on the use of collimated light. This enables relatively large surfaces such as the full human body or an automobile body to be measured in single photographs. Correction in subsequent computer processing of camera errors enables non-metric (ordinary) cameras to be used. Grids have been made of stretched, parallel

Figure 3.1 Principle of parallax as employed in stereo-photogrammetry.

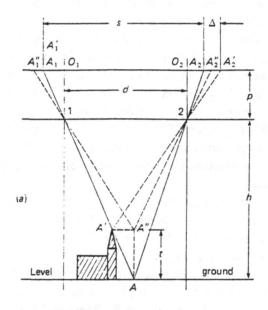

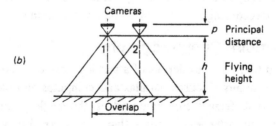

Figure 3.2 Contours of arbitrary surfaces revealed by stereo-photogrammetry. (Prepared by McElhanney Surveying & Engineering Ltd.)

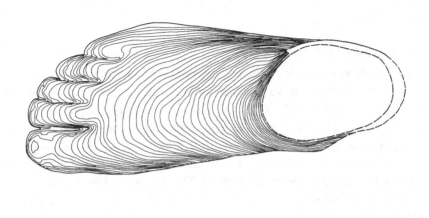

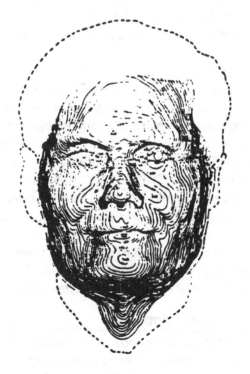

and equally spaced strings which are translated (moved) during photographic exposure to filter out their images, leaving only the fringes. Explanation of the resulting effect lies in Fourier transform analysis. Finally, the shadows have been projected onto the surface, obviating the sometimes awkward necessity of placing a material grid near the surface. We leave further explanations to the published literature.

5. Peri-contourography

This is a development from the conventional shadow-moiré method believed to have originated with the principal author. The shadow-moiré method can be combined with the old technique of peri-photography to produce a plane photograph, replete with moiré fringes, defining the shape of a 'tubular' surface, such as a human limb. Many surface elements which appear to have lateral boundaries can be regarded as portions of 'tubular' surfaces and can also be treated by the periphery method. The representation of a geometrically non-developable surface (such as that of planet Earth) on a plane by chosen transformations to produce the necessary distortions was realised long ago by Mercator in his Projection of the World.

The details of peri-contourography are to be found in Duncan and Mair (1983).

The coordinate data, collected from digitisation of fringes indicating contour levels $z = z_i$, are random in the x, y field. We will show later in Chapter 4 how they may be used via localised parabolic interpolation and surface smoothing to generate a topologically and orthogonally rectangular grid of coordinated points covering the whole field of the surface. This is achieved via the program TRUEPERS. Such coordinates x, y, z may then be called up by programs of the POLYHEDRAL NC® system and used to compute CLD points for tools traversing the surface to machine its likeness. In all of this operator intervention is minimised since the programs involved can be concatenated.

Examples of the use of random point data, processed by TRUEPERS and fed to machining programs NEWERSUE and SUPERSUE, are given in the following chapters and in the program documentation.

6. Beam-scanning Methods

The latest modern methods of surface measurement tend to employ laser beams in various configurations for scanning surfaces. At some instant of time, a straight line pencil of laser light is arranged to be incident on a spot on the surface from a source of emission. The lasers used may be gaseous or solid, chips or infra red. A simple system developed by the authors to demonstrate the principle of parallax-detection by sensing lateral displacement due to increase of depth is fully described in the article by Duncan, Wild, and Hoemberg (1984). Here the surface is scanned mechanically by variation of linear position and angular rotation of a measured model, shown here mounted on a spindle, the laser being fixed. Data were collected and stored automatically in a computer memory so that the system had the potential for full automisation.

Several systems capable of single spot measurement have appeared commercially. A fully automated dynamic system has now been developed in which a scanning laser beam generates three-coordinated points of an observed surface at 1.5mm spacing on an orthogonal grid, within a volume of 20cm x 20cm x 20cm, to ± 0.25mm for z, in about 1.25 seconds! Such data have been used to machine a replica of the observed surface by the authors' POLYHEDRAL NC® system. The whole operation is described in outline in Boulanger et al. (1986). (See Figure 8.6 for an example.)

7. Computed Tomography and CAT Scanning

The methods introduced above are all what might be called 'superficial', using that word in its strict etymological sense – measurement from observing points *above* the surface or from outside of any solid enclosed by a surface. Such methods may encounter difficulties such as 'terminators' (as we see one on the crescent of the moon) due to inclined illumination on steep surface-elements. Even more serious difficulties arise with multi-valued surfaces, of which the bones of the human body provide many examples.

A new principle of surface definition which overcomes some of these difficulties has arrived with the invention of Computed Tomography by Sir Godfrey Hounsfield, a Nobel prize winner. This employs the geometrical idea of sectioning a closed surface by imaginary, parallel, closely spaced 'cuts' through the surface with the aid of X-ray scanning and large-scale matrix inversion in a mainframe or mini-computer. Most large hospitals now have such systems for reconstructing and thus inspecting internal organs and bones of the human bodies of live patients without operation as formerly required. The numerical data used for this *pictorial* reconstruction can also be used for machining replicas of those organs and bones. POLYHEDRAL NC®, with its anti-interference provisions, is able to detect and deal with double values and other contortions in the geometric surfaces which are found in anatomy.

Figure 3.3 shows a typical hospital Computed Axial Tomography (CAT) apparatus. To explain briefly its working, assume it is desired to scan or 'slice'

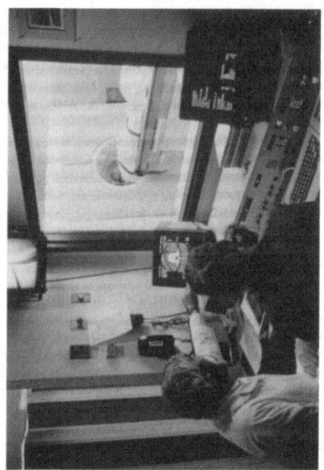

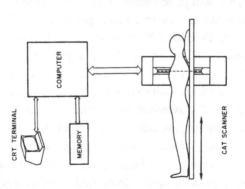

Figure 3.3 A CAT scanner in hospital use.

a live human head in planes normal to an axis through the spine. The patient lies on a couch with his or her head inside an annular aperture as illustrated. An X-ray gun propagates pencils of X-rays through the skull and through the axis of the aperture to a diametrically opposite receptor which detects the strength of each pencil after some of it has been absorbed by all the various body tissues which it has traversed. The beam is arranged to traverse systematically in many incremented angular directions in a plane normal to the axis. When a full rotation of beam direction has been completed, the rotating beam is moved axially to the next, neighbouring plane, usually 1 mm or so away. Hounsfield's discovery or realisation was that different tissues absorb different amounts of X-ray energy. The measure of that capacity for absorption is now the Hounsfield Number.

The composite body traversed by X-rays (the human body) is regarded as composed of many thousands of small (cubical) 'blocks' stacked together to fill all the internal space. Each absorbs X-rays according to its characteristic Hounsfield Number and the emerging beam is thus attenuated by the sum of these elementary absorptions. By recording all the output energies for all angular positions of the beam in each plane examined, a large computer can deduce the Hounsfield Numbers of all the small cubical elements by matrix inversion.

This process is indicated in an extremely simplistic way in Figure 3.4 with attached equations and solutions.

When Hounsfield Numbers are known the boundaries of bones, for instance, being dense and strong absorbers compared with surrounding tissue, can be deduced in a coordinate frame of reference. This detection of Hounsfield Numbers differences is so sensitive that boundaries between two soft tissues of only slightly different Hounsfield Number can be detected; and so a kidney, for instance, can be 'extracted' and imaged for examination.

The coordinates involved in image reconstruction can also be used for replication by machining of bones and organs, for machining bone replacement prostheses. This was done in producing the skull bone shown in Figure 3.5 from sections similar to that shown in Figure 3.6 obtained by CAT scanning the skull of a cadaver. A special version of program NEWERSUE was used for machining. This was modified to suit cylindrical coordinates and to allow for additional interferences in very deep cavities arising from the use of such coordinates.

The latest method of scanning – Nuclear Magnetic Resonance (NMR) – is already here but very expensive. It avoids the use of X-rays, which some

Figure 3.4 A simplistic explanation of the mode of working of computed tomography. (From *Technolgy West*, Winter 1982/1983, pp. 25-26.)

The basis of CAT Scanning simply explained

by J.P. Duncan

The CAT scanner was developed to its final state by G. Housfield, Nobel Prize-winner, and the capacities of elements of material to absorb X-rays are measured by **Hounsfield units**. A "dense" material (bone) may have a greater Hounsfield number than a less dense (flesh). A bone-boundary might then be detected at a point where the Hounsfield number changes abruptly, either up or down. There are now six such machines in the Vancouver area hospitals*. They are capable of defining the shapes of internal organs and bones of the living human body without surgery.

Suppose a region of space is divided into 360^3 cubes each of 0.8mm side (this is one actual division used). Now suppose that a composite biological structure (a human body) is placed in such an "imaginary" region of space and that "pencils" of X-rays are systematically passed through the space in several azimuthal directions. In each such direction the total absorption — sum of Hounsfield numbers can be *measured* and a large set of such measurements made. The problem then is to determine the Hounsfield number of each small element of path traversed by all or any of the many pencils directed through them in a variety of directions.

The resolution of this problem requires matrix inversion of types illustrated below.

Though square are drawn to represent 0.8 mm deep 'slices' through the 360^3 of 0.8 mm cubes, think of each such cube as a small circular cylinder inscribed within the square — a "cell".

UBC, VGH, Royal Columbian St Paul's, B.C. Cancer Control, Lions Gate.

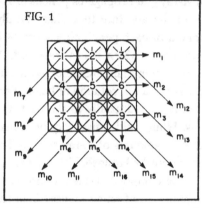

FIG. 1

Suppose the Hounsfield numbers of each *cell* are denoted by X_i, i = 1, 2, 3, ... 9. The m's are the total measured absorptions. Then the following apply:

$$x_1 + x_2 + x_3 = m_1$$
$$x_4 + x_5 + x_6 = m_2$$
$$x_7 + x_8 + x_9 = m_3$$
$$x_1 + x_4 + x_7 = m_4$$
$$x_2 + x_5 + x_8 = m_5$$
$$x_3 + x_6 + x_9 = m_6$$
$$x_1 = m_7$$
$$x_2 + x_4 = m_8$$
$$x_3 + x_5 + x_7 = m_9$$
$$x_6 + x_8 = m_{10}$$
$$x_9 = m_{11}$$
$$x_3 = m_{12}$$
$$x_2 + x_6 = m_{13}$$
$$x_1 + x_2 + x_9 = m_{14}$$
$$x_4 + x_8 = m_{15}$$
$$x_7 = m_{16}$$

Also
$$m_1 + m_2 + m_3 = m_4 + m_5 + m_6$$
$$m_8 + m_{10} = m_{13} + m_{15}$$
and other dependencies may be written. By simple algebra, since
$$x_1 = m_1, x_3 = m_{12}, x_9 = m_{11}$$
and $x_7 = m_{16}$ are known directly from measurements, the remaining unknown x's follow as:
$$x_2 = m_8 - m_4 + m_7 + m_{16}$$
$$x_4 = m_4 - m_7 - m_{16}$$
$$x_5 = m_9 - m_{16} - m_{12}$$
$$x_6 = m_6 - m_{12} - m_{11}$$
$$x_8 = m_{10} - m_{13} + m_8 - m_4 + ...$$
$$.... + m_7 + m_{16}$$

Figure 3.5 Photograph of a machined model of a skull bone. (Duncan, 1982 for CEMAX Inc.)

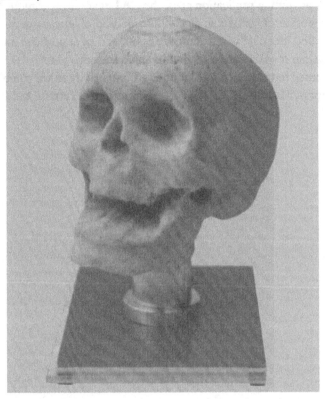

Figure 3.6 CAT scanned sections of a skull bone used in producing the model shown in Figure 3.5.

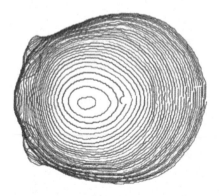

consider undesirable as far as patients are concerned, but has a greater potential for internal examination of the body than current X-ray machines.

As far as routine replication goes, the CAT scanning technique has introduced the notion of sectioning as another method of surface definition. Sections are often identical with contours as obtained by superficial methods but the section can reveal details of a multi-valued surface.

We return to the notion of section in Chapter 6 where we combine it with the concept of Mercator Projection as the basis of the Tetrahedral Concept.

Chapter 4

GRAPHICAL PRESENTATION

1. Introduction

It is universal practice to view three dimensional configurations of surface-bound objects on the screen of a computer terminal as a projection in two dimensions of the linked points of the object defined in three dimensions. This art is now so advanced that it is commonplace to see coloured and shaded two dimensional perspective views or images of objects reconstructed from coordinated surface-point data and seen from chosen viewpoints, virtually instantaneously.

Black and white (monochrome) 'wire frame' images of polyhedra, with or without hidden line removal, are nevertheless of value to see if an algorithm is working or to detect a flaw or error in computations.

A realistic three dimensional image of an object can also be constructed physiologically by having a pair of human eyes view two suitably prepared images to give a stereoscopic perception of depth.

These matters have been reviewed in some detail in Duncan and Mair (1983). In POLYHEDRAL NC® we use central perspective projection in a direct, point-by-point manner to view geometrical configurations on a computer terminal screen. The program TRUEPERS containing a plotting section which, for convenience, we have separated out as program PLOT3D provides all the necessary transformations, profiles and contours normally required and in addition has the capacity to remove hidden lines. We describe this at the end of this chapter and (in technical and operational detail) in its documentation. But first we review briefly some of the basic and elementary processes of computer graphics as far as they are involved in POLYHEDRAL NC®.

2. Dürer's Perspectiva Naturalis

In the sixteenth century, Dürer, the artist, demonstrated the nature of a 'picture' in a framed plane as a set of points corresponding with the spatially distributed set of points of a real object located behind a screen – a picture plane – as seen from a point on the opposite side. With this geometric projective arrangement, a line of sight from an observing eye to an object point can be represented by a taut string. That object-point will appear to be located in the picture plane at the point where the string passes through it.

If two sets of points are drawn in the picture plane, each set corresponding to the view from each eye, the human brain can physiologically reconstruct a realistic image of the real object. A device is necessary to force the brain to receive the appropriate set of points plotted in the picture plane. These devices enable a viewer to see an image of a point-defined object apparently 'floating in space'. Even a defined object which, so far, has had no physical existence may be so viewed.

The above direct or natural method of drawing central perspective images called *Perspectiva Naturalis* contrasts with a derivative method, popularised by Francesca for use by artists, called *Perspectiva Artificialis*.

3. Natural Perspective Transformation Formulae

In Figure 4.1 a gabled building is shown placed in $x\,y\,z$ coordinate space behind a picture plane. Straight lines trace rays from significant object points (corners) to an eye position on the opposite side of the picture plane which in our context will be the screen of a computer terminal. Note that here we use central perspective to explain central perspective which shows that we rely heavily on an intuitive ability in most people to appreciate this three- to two-dimensional transformation. For comparison we show the same 'picture' constructed by way of Dürer's Perspectiva Naturalis, using an elementary computer program.

The transformation is simple and as follows:

$$x_S = x_P + y_P(x_E - x_P)/(y_P - y_E)$$
$$z_S = z_P + y_P(z_E - z_P)/(y_P - y_E)$$

By changing the coordinate values of point E (eye position) we get different views of the object. By letting z_E move to minus infinity we get an orthographic projection, so familiar to engineers.

4. Homogeneous Transformation

If it is desired to keep the point of observation, E, at a fixed position but to move the object in the (imaginary) space behind the picture plane, we must apply in one form or another homogeneous transformation. This translates the origin and rotates the orthogonal axes simultaneously moving the points

Figure 4.1 Perspectiva naturalis and perspectiva artificialis.

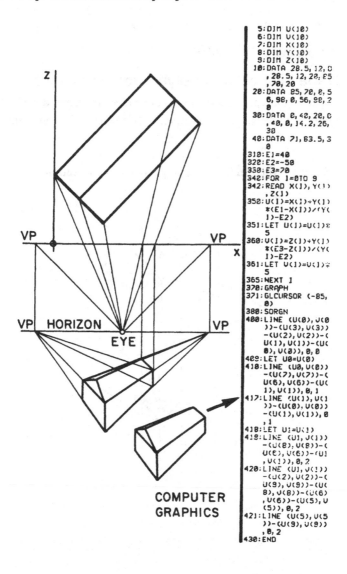

```
5:DIM U(10)
6:DIM V(10)
7:DIM X(10)
8:DIM Y(10)
9:DIM Z(10)
10:DATA 28.5,12,0
,28.5,12,28;85
,70,28
20:DATA 85,70,0,5
6,98,0,56,98,2
8
30:DATA 0,42,28,0
,40,0,14.2,26,
30
40:DATA 71,63.5,3
0
310:E1=40
320:E2=-50
330:E3=70
342:FOR I=0TO 9
342:READ X(I),Y(I)
,Z(I)
350:U(I)=X(I)+Y(I)
*(E1-X(I))/(Y(
I)-E2)
351:LET U(I)=U(I)*
5
360:V(I)=Z(I)+Y(I)
*(E3-Z(I))/(Y(
I)-E2)
361:LET V(I)=V(I)*
5
365:NEXT I
370:GRAPH
371:GLCURSOR (-85,
0)
380:SORGN
400:LINE (U(0),V(0
))-(U(3),V(3))
-(U(2),V(2))-(
U(1),V(1))-(U(
0),V(0)),0,0
402:LET U0=U(0)
410:LINE (U0,V(0))
-(U(7),V(7))-(
U(6),V(6))-(U(
1),V(1)),0,1
417:LINE (U(1),V(1
))-(U(0),V(0))
-(U(1),V(1)),0
,1
418:LET U1=U(1)
419:LINE (U1,V(1))
-(U(8),V(8))-(
U(6),V(6))-(U1
,V(1)),0,2
420:LINE (U1,V(1))
-(U(2),V(2))-(
U(9),V(9))-(U(
8),V(8))-(U(6)
,V(6))-(U(5),V
(5)),0,2
421:LINE (U(5),V(5
))-(U(9),V(9))
,0,2
430:END
```

of the defined object to new positions behind the picture plane. Sometimes it is preferred to translate and rotate separately.

Homogeneous transformation may be applied easily by a computer to all of many points whose coordinates are held in a data file. For various reasons this is an operation frequently required and routinely applied.

These transformations are performed transparently by programs of the POLYHEDRAL NC® system.

In general they represent

$$R_{12} \cdot \rho \; + \quad d \quad + 1 \; = r$$

Rotation + Translation + 1   = Result

5. Perspective View of Polyhedron and Facet

Figure 4.2 shows a typical facet and its normal at its centroid viewed perspectively. There are several other examples of this representation in this book.

Figure 4.2 A polyhedral facet in perspective representation.

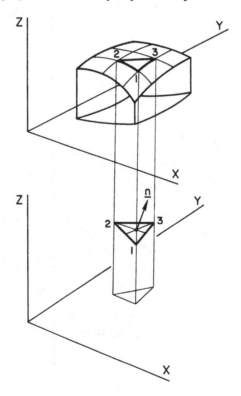

6. The Program TRUEPERS

The authors have used this program since 1971 to prepare and present data for both arbitrary and analytical surfaces defined and presented in terms of dense, surface-point coordinate data. The program was devised by Halstead et al. at the Department of Energy, Mines and Resources, Ottawa, Canada. It was used with permission by the principal author until 1986 and then released to the public.

The program receives random point data (x, y, z) for relatively closely spaced points of an arbitrary surface in a 'field' overlaid with an orthogonal grid of rows and columns intersecting at nodes as in Figure 1.1. Such points would come naturally and very directly from a standard marine chart complete with soundings. They could also come from aerial surveys of landscapes or from mechanical or optical measurements of accessible objects.

There are two main parts of the program: a part originally called ZGRID, which deals with interpolation and smoothing, and another, which we now call PLOT3D, which deals with all the geometric manipulations and computer graphics. We have now separated these two sections of the original program, calling them TRUEPERS and PLOT3D.

The data, which may be listed in any order, are first interpolated locally and parabolically to values at a chosen geometrically orthogonal grid of nodes. A spline-like surface, with a stiffness K chosen by the operator to ensure a desired degree of smoothing, is then fitted by an iteration process until, ultimately, a set of (x, y, z) values at the nodes is determined. The spacing of these nodes has lately been 1.5mm but in many cases has been about 2.5mm.

With interpolated surface-points so determined, preparation can begin for plotting on a terminal screen. Central perspective transformation as described above is used to compute the apparent position on the terminal screen from a selected viewpoint of every surface-point. When these are plotted and linked by straight lines in two orthogonal directions, a set of orthogonal profiles of the interpolated surface is plotted. This has the appearance of a net in true perspective; hence the name, TRUEPERS.

Many alternative views of the interpolated surface can be had by varying the position of the viewpoint. In all cases, hidden lines are removed. This is a relatively lengthy computational process. Optionally, contours of the interpolated surface with reference to the xy coordinate plane can be extracted and plotted by the program.

The well ordered coordinated points of a bi-parametric surface-patch (such as a Bézier or proportionally developed surface within a topologically rectangular but geometrically curvilinear array in the *xy* plane) can be treated as random and interpolated to an orthogonal grid by TRUEPERS and plotted by PLOT3D. Thus these programs have a transforming capacity.

It should also be pointed out that the interpolated surface as ultimately derived passes through the given random points even though these are moved initially to the nearest grid points to start the interpolating process.

Thus TRUEPERS and PLOT3D perform between them many functions concerning surface development and representation by computer graphics. More details of their mode of working are given with the accompanying documentation which also explains how to call up and operate the programs with reference to examples.

Further explanations are given in Duncan and Mair (1983).

Chapter 5

PRISMATIC SURFACES AND TWO-AND-A-HALF-D MACHINING

1. Introduction

Though this work is mainly concerned with arbitrary, unusual, and difficult surfaces requiring machining, we discuss briefly the most usual surfaces encountered in engineering practice – prismatic surfaces. These are frequently assembled in a 'stack', each determined by a contour at a series of levels. Hence the description, $2\frac{1}{2}$ D, which implies two dimensional machining at each level, the $\frac{1}{2}$ referring to the occasional shift required to go from one level to the next. It is commonly estimated that 80% of CAD CAM is concerned with this type of surface definition and machining.

There are many systems, including some 'homegrown' ones of early date of the authors, for completing, on an interactive basis, plan view (x, y) circuits of circular arcs and straight lines between given control points. Many are purely graphical and geometrical and most have very comprehensive capacities for transformation, sectioning, dimensioning, etc., as required in engineering drawing. Some go further in deducing Cutter Location Data (CLD), as required to machine closed circuit outlines at each of a series of levels; these have $2\frac{1}{2}$ D capacity in both graphics and machining.

If the circuits referred to above are curves of general functional form and are not parts of circular arcs or straight lines, the representation of such forms by many-sided polygons becomes necessary. These can then be generated by tool motion along chords or secants of that curve.

In this chapter we deal only with this type of general prismatic surface and with the complications that may arise in practical cases; with the interference which may occur and its avoidance by means of the program CAL-TOOLP, part of POLYHEDRAL NC®.

51

2. Prismatic Surfaces

A prismatic surface is described or generated by the continuous movement of a straight line parallel to itself (of constant direction in space) while it passes through and traces out a space-curve, often closed. See Figure 5.1 . Frequently the line is parallel to a z axis and the curve is defined in an $x\,y$ coordinate plane, parallel to the plane of $z=0$, as we shall assume in further discussion and examples.

A 'stack' of such surfaces defined and machined between a series of levels, $z = z_1, z_2, z_3$, etc., constitutes what we call a terraced surface, the governing curve of each cylindrical element being a contour of some more general global surface as illustrated in Figure 5.2 . Such terraced models are preferred by some as physical approximations to smooth, global models of what are often called *sculptured surfaces*.

3. Analytical Prismatic Surfaces

An analytical, prismatic surface may have an assigned equation or may be piece-wise fitted through some digitised points by Hermitian or some other curve-fitting process. In such cases that it may be possible to determine the tangents and normals to such curves by differentiation. Tools should not have radii of curvature larger than the local radii of curvature of the curve-elements being cut from offset positions of the tool axis. In Cartesian coordinates the curvature, K, is given by:

$$K = y'' / ((y')^2 + 1)^{1.5}$$

Figure 5.1 A typical cylindrical surface at a given contour level.

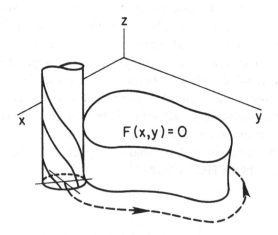

where coordinates are x and y and prime means differentiation with respect to x.

In polar coordinates: $K = (r^2 - r.r' + 2(r')^2)/(r^2 + (r')^2)^{1.5}$

Figure 5.2 The geometry of terracing.

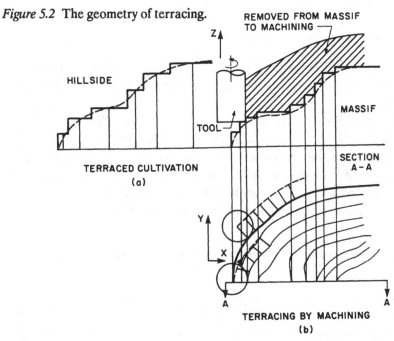

Figure 5.3 Derivation of the Joukowski airfoil shape by conformal transformation of a circle.

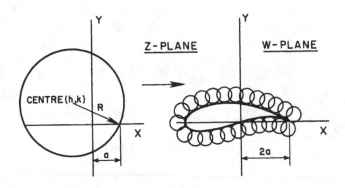

where r, θ are polar coordinates and prime means differentiation with respect to θ.

Some examples where these differentiation processes might be applied are as follows. The well known Joukowski theoretical airfoil as shown in Figures 5.3 and 5.4 has the Cartesian equation:

$$y = k \pm \sqrt{(k^2 - (x^2 - 2xh + 2ah - a^2))}$$

obtained by a well known conformal transformation. This might be used to determine CLD; or polygonal machining, as follows, could be used and may be simpler.

The unusual curve, called by Lockwood (1961) 'Burleigh's Oval', Figure 5.5, with $e^2 = {}^1\!/_2$, gives a curve which we have called 'Burleigh's Fish'. This is interesting and instructive because it has an intersection where the 'tail' joins the 'body' and has cusps at the tips of the 'tail'. To machine this outline, changes in tool path direction at the points mentioned have to be computed (best by iteration) from the equation of the curve which is:

Figure 5.4 A cascade of Joukowski airfoil shape as cylindrical surfaces.

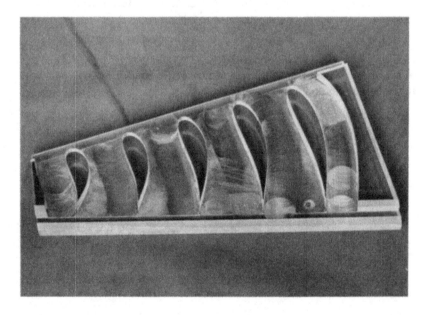

$$(2x^2 + y^2)^2 - 2\sqrt{((ax) \times (2x^2 - 3y^2) + 2a^2(y^2 - x^2))} = 0$$

This example leads us to a general approach to the machining of any prismatic surface along whose section closely spaced coordinated points are known, either by digitisation or by computation from an equation. The continuous curve can be represented approximately by a many-sided polygon formed by joining such points with straight lines. The prismatic surface can then be machined adequately as a prismatic polygonal surface. Our program in the POLYHEDRAL NC® system for doing this with full anti-interference protection is called CALTOOLP, written by K. K. Law, based on TOOLP of the POLYHEDRAL NC® system.

4. Terraced Surfaces

In these a ledge is cut in a block of raw material at the level of each of many contours defining the surface as illustrated in Figure 5.2 . The tool used is a side and flat-ended milling cutter. The resulting model is terraced with many steps as commonly seen on cultivated mountain sides. Figure 5.6 shows examples. The ledge between the contours is generally of variable width and may often be wider than the tool diameter. Thus to machine the ledge everywhere, it is often necessary for the tool to make several circuits of a con-

Figure 5.5 Burleigh's 'fish' with evolutes.

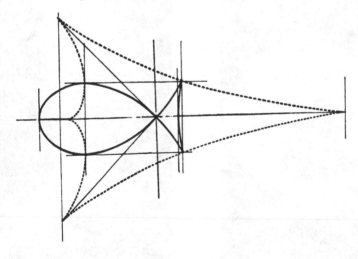

tour at several offsets, each greater than the last.

The tool centre (projected point-view of its vertical axis) must be directed to trace the linear elements of the approximating polygon referred to above. Figure 5.7 shows how intersections of these offset linear elements must be found to limit the tool travel along each element and so avoid removing required material.

However, in machining highly contorted contours such as those of a human face shown in Figure 5.6 the tool, if unchecked, may remove material at points ahead or behind its present position while cutting correctly at its present position in its path of progress. We call this interference of which four types have been identified.

5. Interference in Contour Machining

Inadvertent removal of material by interference is prevented during the pre-calculation of Cutter Location Data (CLD) by program CALTOOLP in the POLYHEDRAL NC® system.

There are at least four types of interference.

(i) *Small radius of curvature interference*. The most obvious interference

Figure 5.6 Terraced models of a face and a radius bone (of the arm).

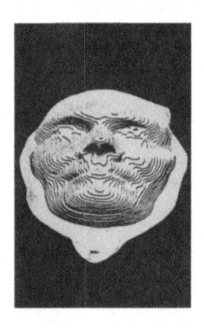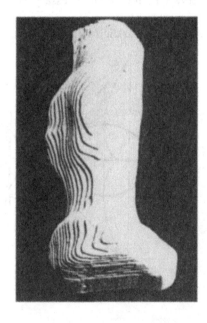

is when the cutter radius is greater than the local 'average' radius of curvature of the digitised surface as shown in Figure 5.8 . In this case the tool must be stopped in its progress along the chord P_{i-1} P_i before it invades any other polygonal face either before or after the one it is on.

(ii) *Small corner interference*. This is illustrated by Figure 5.9 which shows the 'natural' intersection points of the offset polygonal paths. Obviously the tool must not visit point T_i but, rather, the point of intersection of $T_{i-1}T_i$ and $T_{i+1}T_{i+2}$, thus by-passing 'the small corner'. The only way to get into that corner is to use a smaller tool.

Figure 5.7 Tool path offset from polygonal elements fitting a curve.

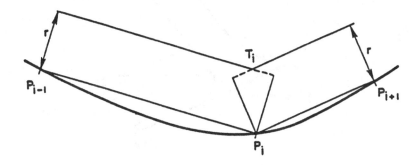

Figure 5.8 Small radius of curvature interference.

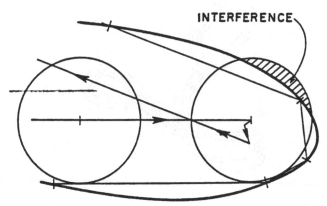

(iii) *Narrow gap interference.* Contours may be as shown in Figure 5.10 . In passing along and cutting a contour on one side of the gap at appropriate offset, the tool may, obviously, cut away required material on the other side. Thus when the tool ultimately reaches that other side, there be not be any material to cut away! Also, if the tool were correctly placed to cut that second

Figure 5.9 Small corner interference.

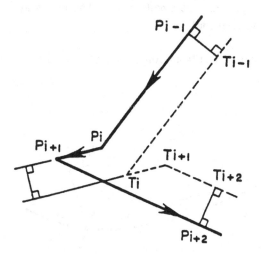

Figure 5.10 Narrow gap interference.

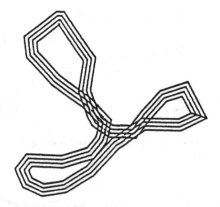

side, it would remove material from the side of the contour that it had cut correctly when it visited it earlier. Again, the only way to cut both sides of a gap correctly is to use a tool that can pass through. Failing that, the program CALTOOLP automatically 'lifts' the tool out of cutting range as it traverses elements of path where interference would occur at points either ahead or behind the present tool position. These elements of path where the tool has been 'lifted' are shown dotted.

(iv) *Contours with branches.* An example of this would be a 'mountain within a valley' as shown in Figure 5.11 . The 'mountain' is represented by closed 'ring' contours; the valley by 'strands'. If we are following a 'ring' on the outside with a tool to machine terraces on the 'mountain' we may cut into the 'valley'. The 'strands' of the 'valley' as shown may be parts of two other 'mountains' – they may close into rings representing two 'mountains' adjacent to the first – or they may join each other to form a 'ring' completely surrounding the first 'mountain'. In the latter case we may have two closed contours at the same level within reach of each other. So we see here potential for interference between branches of contours of the same level. We cannot afford to concentrate on only one branch, examining it both before and after the present position: we must also notice other branches at the same level.

Figure 5.11 Interference with multiply connected contours.

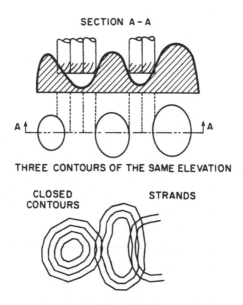

SECTION A – A

THREE CONTOURS OF THE SAME ELEVATION

CLOSED
CONTOURS   STRANDS

Program CALTOOLP examines and corrects for possible interferences of this type.

6. Application

CALTOOLP was derived from an earlier program called TOOLP in POLYHEDRAL NC®. It directs the tool to follow a digitised contour. If the contours as given do not cross (in which case they would be erroneous) no interference can occur.

CALTOOLP has been used mostly for machining contoured landscapes and medical models such as bones measured by sectioning and contouring, the contours being derived from aerial survey maps, shadow-moiré or CAT scanning via digitisation. In other cases where a surface has an equation, it is sometimes more convenient to compute points along contours and to use CALTOOLP rather than to differentiate awkward functions.

More is said about the operation and management of CALTOOLP in the documentation.

Chapter 6

CLOSED POLYHEDRA AND MERCATOR PROJECTION

1. Introduction

In previous chapters the many-faced, irregular polyhedron has been visualised as a partially bounded surface having a lateral boundary in an x, y plane of projection. The sum of the areas of its facets provided an approximation to the total area of the circumscribing continuous surface and the centroid and normal of each facet provided a basis for fixing suitable tool-centre positions for machining replicas.

Many surfaces of practical interest, such as bones or organs of the human body, are completely enclosed by a surface. Further, such surfaces are now being measured and defined by many closely spaced sections, as in CAT scanning, described earlier. Single valued surfaces may often be defined in terms of complete contours of two conceptual 'quasi-hemispheres' measured by superficial methods. Also the method of peri-contourography produces data for tubular surfaces which can readily be closed at the ends as will be shown.

For *closed* polyhedra, the volumetric properties of the volume enclosed can be found by joining each vertex of the polyhedron to an internal point (subject to some conditions discussed below) thus forming a complete packing of that volume with tetrahedra of which the facets may be regarded as the bases and the common internal point as vertex of all of them.

Such tetrahedra then become the elements of a TETRAHEDRAL CONCEPT, a general theoretical concept which 'lifts' POLYHEDRAL NC® by one dimension.

2. The Arbitrary Solid

A solid of arbitrary shape is shown in Figure 6.1 . If the density of this solid is constant, it is homogeneous; if not, its density is a function of coordinates x, y, z and it is non-homogeneous. For simplicity in explanation (but not as a limitation on the scope of this concept) we will assume a uniform density of 1 as far as mass is concerned.

The solid may be convex or concave. If convex, a straight line drawn in space from a point on the enclosing surface will intersect another region of the surface once only; if concave it may intersect it at two or more other points as illustrated in Figure 6.2.

In the case of a convex solid, an internal point D anywhere within the bounding surface subtends, internally, small curvilinear elements of surface everywhere on the surface. With a concave or partially concave surface, only points D within the shaded region bounded by a surface formed by tangential

Figure 6.1 A solid of arbitrary shape bounded by a closed surface together with its 'Mercator Projection'.

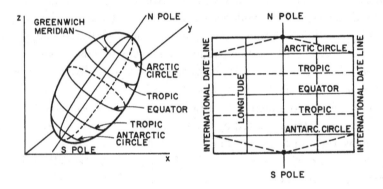

Figure 6.2 Convex and concave closed surfaces.

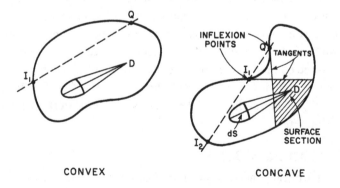

planes at curves of inflexion, as illustrated two dimensionally, subtend such elements internally.

Truly polyhedral solids as in Figure 6.3 can be treated either as an assembly of elementary units or as a Mercator Projection (Figures 6.1 and 6.3).

3. Polyhedral Representation by Mercator Projection

Consider an arbitrary solid enveloped by either a convex (or a partially concave surface) as in Figure 6.1. Such a surface may be overlaid with arbitrary lines of 'latitude' and 'longitude' whose analogy with those of planet Earth will be obvious and whose alternative representation by a 'Mercator Projection' will be clear and easily appreciated from the figure. Intersections of specific lines of 'latitude' and 'longitude' have Cartesian coordinates x, y, z and can be taken as vertices of an inscribed/circumscribed polyhedron approximating in surface area and volume to the continuous arbitrary solid bounded by its curved surface. Such coordinates are often computed on a fine scale, as in medicine via Computer Aided Tomography (CAT scan). Alternatively, they may be obtained by automatic digitisation of sets of experimentally determined contours or can be computed for solids having known analytical bounding-surface equations.

Figure 6.3 A surface compounded from two polyhedra viewed as a 'global' surface together with its Mercator Projection.

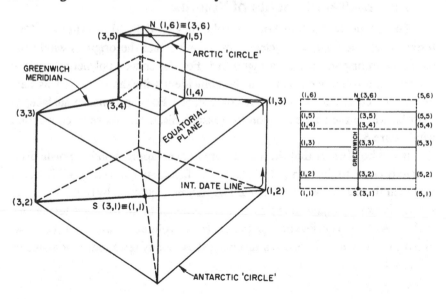

These coordinates, x, y, z, of the vertices of a real or approximating surface can be computer-stored for recall and processing at the 'nodes' of an indicial grid — a 'Mercator Projection' in Figure 6.1 .

Note that all nodes in row 1 have identical coordinates x_S, y_S, z_S, those of the 'South Pole'. Likewise, all in the Nth row have coordinates x_N, y_N, z_N, those of the 'North Pole'.

All facets of the polyhedral representation of the original surface can be fully defined geometrically in terms of coordinates stored at indicial locations I, J of the 'Mercator Projection'. For instance, in each 'cell', triangular plane facets ABC and ACD as in Figure 6.1 can be defined geometrically in terms of standard analytical coordinate geometry.

Note that the 'ends' of the solid are formed by joining North and South Poles by straight lines to points on the 'Arctic and Antarctic Circles' respectively.

Though curvilinear bounding surfaces and their approximate representation by polyhedra have been visualised so far, a real polyhedron, such as a right parallelepiped or a compound of regular faceted solids as shown in Figure 6.3 may be the subject of interest. Such surfaces can be represented accurately by a 'Mercator Projection', either as a solid bounded by a single surface or by several primary surfaces assembled to make up the whole. The former concept can be understood by noting the correspondence in Figure 6.3.

4. Tetrahedral Elements of Volume

Though in the final analysis the solid may be defined anywhere in Cartesian space, we assume for clearer explanation that the origin is within the closed bounding surface as in Figure 6.4. For the convex polyhedral surface we choose a reference point D to be coincident with the origin. (This particular choice is not essential but convenient.) For the concave surface, D must be within the shaded region for reasons discussed above in connection with Figure 6.2 .

It will be obvious that the surface area of the approximating polyhedron is exactly equal to the sum of the areas of all facets such as ABC, ACD represented in the I, J array of Figure 6.5. The volume contained by that polyhedral surface is exactly equal to the sum of the volumes of all irregular tetrahedra such as ABCD. The location of the centre of volume or mass can be computed by summation processes involving these many tetrahedral volumes as will now be shown.

Chapter 6

CLOSED POLYHEDRA AND MERCATOR PROJECTION

1. Introduction

In previous chapters the many-faced, irregular polyhedron has been visualised as a partially bounded surface having a lateral boundary in an x, y plane of projection. The sum of the areas of its facets provided an approximation to the total area of the circumscribing continuous surface and the centroid and normal of each facet provided a basis for fixing suitable tool-centre positions for machining replicas.

Many surfaces of practical interest, such as bones or organs of the human body, are completely enclosed by a surface. Further, such surfaces are now being measured and defined by many closely spaced sections, as in CAT scanning, described earlier. Single valued surfaces may often be defined in terms of complete contours of two conceptual 'quasi-hemispheres' measured by superficial methods. Also the method of peri-contourography produces data for tubular surfaces which can readily be closed at the ends as will be shown.

For *closed* polyhedra, the volumetric properties of the volume enclosed can be found by joining each vertex of the polyhedron to an internal point (subject to some conditions discussed below) thus forming a complete packing of that volume with tetrahedra of which the facets may be regarded as the bases and the common internal point as vertex of all of them.

Such tetrahedra then become the elements of a TETRAHEDRAL CONCEPT, a general theoretical concept which 'lifts' POLYHEDRAL NC® by one dimension.

2. The Arbitrary Solid

A solid of arbitrary shape is shown in Figure 6.1 . If the density of this solid is constant, it is homogeneous; if not, its density is a function of coordinates x, y, z and it is non-homogeneous. For simplicity in explanation (but not as a where s = the semi-perimeter of the triangle. The lengths a, b, c equal to AB, BC, CA are each found by use of Pythagoras' theorem and the coordinates of A, B, C. (The relative merits of calculating areas by this and alternative methods involves tracing the number and nature of computer operations involved in each method.) The area of the whole polyhedral surface is found by adding the areas of all constituent facets.

The volume of a tetrahedron such as ABCD is given by a well known standard result, the absolute value of the 24 term determinant:

$$V = (1/6) \times \begin{vmatrix} x_A & y_A & z_A & 1 \\ x_B & y_B & z_B & 1 \\ x_C & y_C & z_C & 1 \\ x_D & y_D & z_D & 1 \end{vmatrix}$$

where x_A, y_A, z_A, etc., are the coordinates of the tetrahedral vertices A, etc.

The total volume of the solid enclosed by the approximating polyhedron is exactly equal to the sum of the volumes of all such tetrahedra as ABCD which nest exactly within the polyhedron, as is shown by example in the documentation of the program VCAM.

To locate the centroid of the solid, the sum of the moments of volume (or of mass) of all tetrahedra about coordinate axes of x, y, z may be found and divided by the volume itself as found from the above determinant to yield the coordinates x_P, y_P, z_P of the volume (or mass) centroid of the whole solid at point P as shown in Figure 6.4.

The moment of volume (or mass moment) of the typical individual tetrahedra can be found by reference to Figures 6.6 and 6.7. The centroid of typical plane triangle ABC is at G, the intersection point of medians. The volumetric (or mass) centroid of the typical tetrahedron ABCD is at T at a distance from the origin (reference point) equal to $(3/4) \times DG$.

The direction cosines α, β, γ of DG are readily computed. Then the moments of volume (mass moments) of the tetrahedron about the axes x, y, z are:

$$m_x = v.(DT)\,\alpha, \qquad m_y = v.(DT)\,\beta, \qquad m_z = v.(DT)\,\gamma$$

where v is the volume of the elementary tetrahedron.

The total moments of volume (or mass moments) of the whole solid enclosed by the polyhedron are therefore:

$$M_i = \Sigma\, m_i \quad \text{where } i = x, y, z$$

The coordinates of the volume centroid (and mass centroid) of the whole solid are therefore:

$$x = M_x/V, \quad y = M_y/V, \quad z = M_z/V$$

6. Second Moments of Volume and Inertia

Second moments of volume (and of inertia if density is included) are conveniently expressed with respect to the volume or mass centroid as a reference point. The centroid is always within a convex polyhedron as illustrated in Figure 6.4. (If it is not within the shaded region of any concave polyhedron, a special treatment not pursued here is necessary.)

Figure 6.6 Determination of moment of volume of an elementary tetrahedron.

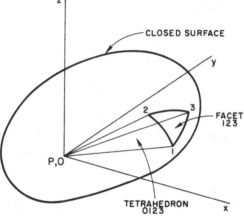

Figure 6.7 Determination of moments and products of inertia about the centroid of a surface-enclosed solid.

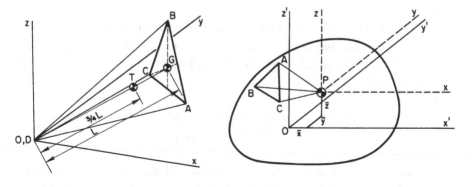

First move the origin of surface definition without rotating the axes to the centroidal location at P as shown in Figure 6.7. The second moments of volume (or of mass) of the whole solid can now be computed by summing the contributions of all the constituent tetrahedra having P as a common vertex. (To simplify writing, we have changed coordinate symbols here: old coordinates are primed, new ones unprimed.)

A cubical element of volume dx, dy, dz within a single tetrahedron of density distribution $\rho = \rho(x, y, z)$ has a second mass moment of inertia about the x axis of:

$$J_x = \iiint (y^2 + z^2) \cdot \rho(x, y, z)\, dx\, dy\, dz$$

and has products of inertia of:

$$J_{xy} = \iiint (x\,y) \cdot \rho(x, y, z)\, dx\, dy\, dz$$
$$J_{zx} = \iiint (z\,x) \cdot \rho(x, y, z)\, dx\, dy\, dz$$

where we allow $\rho = \rho(x, y, z)$ to vary as a known function of x, y, z.

(Such variation of density is routinely revealed on a very fine scale but in three dimensional gridded form by Computed Tomography (CT scan).) For homogeneous bodies, ρ is constant and we will continue by assuming homogeneity with $\rho = 1$.

These triple integrals can be evaluated for any space function within a tetrahedron by means of an algorithm having its origin in finite element analysis. We append an extract from *The Chinese Mathematical Handbook of Standard Results* giving this algorithm amongst many other similar ones relating to other solids.

The values of the triple integrals for a tetrahedron within which a function $f(x, y, z)$ exists and applies is given by the values of the function at (a) the four vertices of the tetrahedron and (b) the four facial centroids of the four faces as follows:

$$\iiint f(x, y, z)\, dx\, dy\, dz = \Sigma\, v \times ((1/40) \cdot f(x_i, y_i, z_i)) +$$
$$\Sigma\, v \times ((9/40) \cdot f(x_j, y_j, z_j))$$

where $x_i, y_i, z_i, i = 1, 2, 3, 4$, are the coordinates of the four vertices and $x_j, y_j, z_j, j = 1, 2, 3, 4$, are the coordinates of the four facial centroids, and v is the volume of the (single) tetrahedron as computed above.

For a homogeneous body, $f(x, y, z)$ in the present case is $(y^2 + z^2)$ for calculating J_x and is (xy) and (xz) for J_{xy} and J_{xz} respectively. Similar expressions, as will be obvious by analogy, apply to J_y, J_z, and J_{yz}. Note that $J_{yz} = J_{zy}$, etc. If the form of the density function $\rho(x, y, z)$ had been known, the function under the triple integral sign would change but otherwise the evaluation would be by the same algorithm.

Thus the second moments of volume and mass of every constituent tetrahedron about axes parallel to the original through the centroid at P can be computed in a fundamental way. The moments and products of the whole solid can then be found by direct summation. These global properties have been denoted by I_{xx}, I_{yy}, I_{zz}, I_{xy}, I_{yz}, I_{zx} to distinguish them from the elementary tetrahedral values denoted by J_{suffix}.

7. Principal Second Moments of Volume and Mass

The arbitrary directions of the coordinate axes of definition of the solid shown in Figure 6.5 are obviously not those of the principal axes.

Since I_{xx}, I_{yy}, I_{zz} and $I_{xy} = I_{yx}$, $I_{yz} = I_{zy}$, $I_{zx} = I_{xz}$ are all calculable for the whole solid as explained above, the invariants of the second moments can be found as follows:

$$A = I_{xx} + I_{yy} + I_{zz}$$

$$B = I_{xx} \times I_{yy} + I_{yy} \times I_{zz} + I_{zz} \times I_{xx} - I_{xy} \times I_{xy} - I_{yz} \times I_{yz} - I_{zx} \times I_{zx}$$

$$C = I_{xx} \times I_{yy} \times I_{zz} - 2 I_{xy} \times I_{yz} \times I_{zx} - I_{xx} \times I_{yz} \times I_{yz} - I_{yy} \times I_{zx} \times I_{zx} - I_{zz} \times I_{xy} \times I_{xy}$$

The values of the three principal moments are the roots of the cubic equation (all roots must be real for a physical solid):

$$I^3 - A \times I^2 + B \times I - C = 0$$

whose solution is routine.

Let the roots be I_1, I_2, I_3 or I_i, $i = 1, 2, 3$.

For each root I_i,

$$\begin{vmatrix} (I_{xx} - I_i) & -I_{xy} & -I_{xz} \\ -I_{yx} & (I_{yy} - I_i) & -I_{yz} \\ -I_{zx} & -I_{zy} & (I_{zz} - I_i) \end{vmatrix} = 0$$

If l, m, n are direction cosines of any principal axis of the volume or solid having principal second moment I_i, then l, m, n may be found by solving the following four equations, a well known procedure for a two dimensional tensor of which I_i are eigenvalues:

$$\begin{vmatrix} (I_{xx} - I_i).(l) & -I_{xy}.(m) & -I_{xz}.(n) \\ -I_{yx}.(l) & (I_{yy} - I_i).(m) & -I_{yz}.(n) \\ -I_{zx}.(l) & -I_{zy}.(m) & (I_{zz} - I_i).(n) \end{vmatrix} = 0$$

$$l^2 + m^2 + n^2 = 1$$

8. Related Properties

If in the tetrahedron ABCP discussed above the vertex P lies in the plane ABC, the value of the expanded determinant will be zero. This is then the condition that four points lie in a plane. (Three points always do; that is why POLYHEDRAL NC® is based on triangular facets.)

If a point P is enclosed anywhere within a closed, *convex* surface represented by a polyhedron, and is used as a common vertex for a set of tetrahedra based on each and every one of the triangular facets of that polyhedron, the total volume as computed by the TETRAHEDRAL CONCEPT will be the same for all locations of P. If the point P is not enclosed, the calculation will not give a constant answer. This constitutes a test of whether a given point in a coordinate space is inside or outside of an enclosure – does the point 'invade' the enclosed space?

In the analysis of possible collision in a pre-planned robotic path, the essential problem is to see whether one surface-enclosed volume, such as a link of a robotic arm, co-exists in space at any location and time with another, say a workpiece or a workspace. If at least one of two polyhedra involved in a possible collision is finely specified in terms of many vertices, each such vertex can be tested as a point to see if it is within the volumetric space of the other. If so, collision will occur in the proposed pre-planned path. This was one of two methods used by Duncan et al. (1984).

The Tetrahedral Concept can also be readily and specially adapted for determination of the properties of propeller blades.

9. Computer Program VCAM

The program VCAM for computing Volume, Centroid location, Area and Moments and their principal values for an arbitrary solid based on the foregoing ideas is described and its accuracy demonstrated in the documentation

FROM CHINESE MATHEMATICS HANDBOOK

第六章 积分学

§2 多重积分、曲线积分与曲面积分

[三重积分的近似计算公式] (MOMENT OF INERTIA)

$$\iiint_V f(x,y,z)\,dx\,dy\,dz = A_V \sum_{i=1}^n w_i f(x_i,y_i,z_i) + R$$

式中 A_V 对于不同的积分区域 V 选取不同的常数，w_i 是求积系数，R 是余项。

Γ 为球体 S: $x^2+y^2+z^2 \leqslant h^2$, $A_S = \frac{4}{3}\pi h^3$

n	图示	(x_i, y_i, z_i)	w_i	R
7		$(0,0,0)$	$\frac{2}{5}$	
		$(\pm h,0,0)$	$\frac{1}{10}$	$O(h^5)$
		$(0,\pm h,0)$	$\frac{1}{10}$	
		$(0,0,\pm h)$	$\frac{1}{10}$	

Γ 为立方体 C: $|x|\leqslant h, |y|\leqslant h, |z|\leqslant h$, $A_C = 8h^3$

n	图示	(x_i, y_i, z_i)	w_i	R
6		$(\pm h,0,0)$	$\frac{1}{6}$	
		$(0,\pm h,0)$	$\frac{1}{6}$	$O(h^5)$
		$(0,0,\pm h)$	$\frac{1}{6}$	
21		$(0,0,0)$	$\frac{496}{360}$	
		中心到 6 个面的距离减为 $\frac{6}{h}$ 个面的中心	$\frac{128}{360}$	$O(h^5)$
		6 个面的中心	$\frac{8}{360}$	
		8 个顶点	$\frac{5}{360}$	

n	图示	(x_i, y_i, z_i)	w_i	R
42		6 个面的中心	$\frac{91}{450}$	
		12 个棱的中点	$\frac{40}{450}$	$O(h^3)$
		每个面的对角线上到中心距离为 $\frac{h}{2}\sqrt{5}$ 的 4 个点 (共 24 点)	$\frac{16}{450}$	

Ω 为四面体 T. $A_T = V$ 为四面体体积 A_T: VOLUME OF THE TETRAHEDRON

n	图示	(x_i, y_i, z_i)	w_i	R
8		4 个顶点 4 VERTICES	$\frac{1}{40}$	
		4 个面的重心 4 CENTROIDS	$\frac{9}{40}$	
11		T 的重心 T : CENTROID	$\frac{8}{15}$	
		4 个顶点 4 VERTICES	$\frac{1}{60}$	
		6 个棱的中点 6 MID POINTS OF EDGES	$\frac{1}{15}$	

[曲线积分的近似计算公式]

圆周 Γ: $x^2+y^2 = h^2$ 上的曲线积分

$$\int_\Gamma f(x,y)\,ds = \frac{\pi h}{n}\sum_{k=1}^n f\left(h\cos\frac{k\pi}{n}, h\sin\frac{k\pi}{n}\right) + O(h^{2n-1})$$

[曲面积分的近似计算公式]

球面 Ω: $x^2+y^2+z^2 = h^2$ 上的曲面积分

accompanying the VCAM program on disk. It is designed to evaluate the arbitrary body but it has been tested for accuracy against some regular, well known solids such as parallelepipeds and ellipsoids.

The TETRAHEDRAL CONCEPT was a natural outgrowth and extension of many years of development of the POLYHEDRAL NC® system, a data storage system which anticipated the present ready availability of large data storage capacities in personal computers. These two systems are designed to deal with arbitrary (sculptured) surfaces rather than the regular ones.

The authors have subsequently discovered the publication by Timmer and Stern (1980) of which they had no previous knowledge. This paper uses Green's theorem to find the mechanical properties of some regular solids, with solids with surfaces bounded by bi-cubic patches in mind for more general application.

Chapter 7

MACHINING ALGORITHMS

1. Introduction

Once a curved surface has been represented approximately by an irregular polyhedron whose vertices, when projected onto the x, y coordinate plane form a topologically rectangular grid, we may proceed to machine it with tool positions computed by one of two alternative algorithmic methods. These are represented by programs NEWERSUE and SUPERSUE.

NEWERSUE is the latest version of what was originally developed as SUMAIR, the general theory of which was first presented in the proceedings of the conference PROLAMAT 76 published as *Advances in Computer-aided Manufacture*, D. McPherson, Editor, North Holland Publishing Company, 1977. It was also published later in Duncan and Mair (1983). NEWERSUE is a theory of tool positioning which is based on the idea of causing a spherically-ended tool to touch in turn all the plane facets of the polyhedron without interference.

SUPERSUE is a very recent, as yet unpublished development in which the same spherically-ended tool is positioned in turn with its axis centred at each and every node of a topologically and geometrically rectangular grid at such a height in direction z that no vertex of the approximating polyhedron is within the spherical tool-end surface. The method is a non-interfering one with simplified computing and greater speed in machining than with NEWERSUE. It is particularly useful for machining with multiple machining heads. SUPERSUE can also 'cascade' a set of tools as in NEWERSUE.

Both methods aim to deal with the same problem of interference – the inadvertent removal of required material in regions of high curvature – by different algorithms for tool positioning. The latest versions of both programs include a full field check by finite difference calculus of Gaussian expressions for principal curvatures. Tools cannot be of less curvature than these values without interference (unless retraction is applied to avoid it).

When all facets or nodes of a field have been visited by the tool and a surface cut, the errors implicit in each of these approximate machining methods are computed in terms of the material left unremoved above each surface-node.

Three dimensional perspective plots of curvatures and errors at nodes may be displayed on demand.

This chapter discusses the main features of both NEWERSUE and SUPERSUE. Directions for the use of the programs and some of the details are given in the accompanying documentation.

2. The Basics of Program NEWERSUE

This program computes the cutting-tool centre-location (CLD) to ensure that the tool will touch in turn each facet of a polyhedron during a systematic visitation, or be withdrawn appropriately to avoid interference at any facet if actual touching would cause interference to occur. The general idea was outlined in Chapter 1.

Since machining is usually carried out in a machine operating in Cartesian coordinates and needs visualisation in that frame of reference to assist description, we first give such a description and follow later with a concise vector presentation.

Figure 7.1 shows a projected curvilinear grid of m rows and n columns in the xy coordinate plane at whose nodes the height z to a point on a surface is defined. Such a point may be specified by a vector r with Cartesian coordinates.

The polyhedral approximation to a surface consists of many triangular facets such as 123 and it is desired to touch each in turn in a systematic manner with a sphere of radius R representing the cutting end of a milling cutter.

The coordinates of typical vector r, computed or collected as data in various ways as described in Chapters 2, 3, and 4, can be stored for easy recall in a computer random access memory (RAM). The points such as 1, 2, or 3 can be identified for data storage and recalled by specifying row and column in the topologically rectangular grid.

The equation of the plane of the facet 1 2 3 is, in general,

$$ax + by + cz + d = 0$$

Using the coordinates of points 1, 2, and 3, the three-point equation of the plane is expressed by:

$$\begin{vmatrix} x & y & z & 1 \\ x_1 & y_1 & z_1 & 1 \\ x_2 & y_2 & z_2 & 1 \\ x_3 & y_3 & z_3 & 1 \end{vmatrix} = 0$$

From this, as explained in Chapter 2, the coefficients (direction numbers of the normal to the plane) a, b, c can be evaluated, and by normalising, the direction cosines α, β, γ are found. Then the normalised equation of the plane becomes:

$$\alpha x + \beta y + \gamma z + p = 0$$

where in a practical context γ is made positive (by multiplying through by minus 1 if necessary) and p is the signed perpendicular distance of the plane

Figure 7.1 One polyhedral facet with bordering facets projected onto a curvilinear, topologically rectangular grid of projected node-points of a surface.

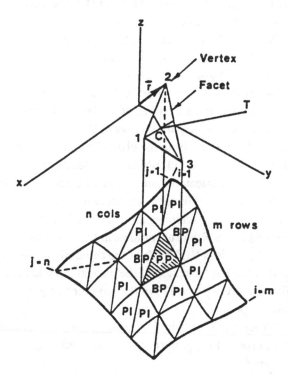

from the origin. The coordinates of the tool-sphere centre-location at a point, T, when touching the centroid, C, of the triangular facet 123 are:

$$x_T = x_C + R \cdot \alpha$$
$$y_T = y_C + R \cdot \beta$$
$$z_T = z_C + R \cdot \gamma$$

The tool-facet relationship is shown clearly in Figure 7.2.

3. Anti-interference Measures in NEWERSUE

The typical facet 123 discussed above is called the principal plane (PP) (of current interest). It is surrounded by three planes having two vertices, that is, one edge, in common with it, and these are called bordering planes (BP). There are nine more planes having one vertex in common with 1, 2, or 3. These are called planes of interest (PI). All of these planes may be within reach of the tool when touching plane 123 at centroid C. In fact, the tool may, in some cases, reach to other planes of interest (PI) which, though near, have no vertices in common with 1, 2, or 3.

Figure 7.2 shows in three dimensional perspective a tool of spherical radius R positioned at T touching a principal plane 123 at C. Three bordering planes, $13V_1$, $32V_3$, and $21V_2$ are also shown.

It is well known from analytical coordinate geometry that substitution of the coordinates of a point in the left hand side of the normalised equation of a plane, as found above for a principal plane, gives the signed value of the perpendicular distance of that point from the plane. Thus, if we first find the vertices of any and every facet, BP or PI, which has at least one vertex within the projected circle of the cylindrical tool-shank, the distance of that vertex from principal plane PP can be found by such substitution.

In a physical situation of machining, the point T will be 'above' the PP. Its distance from PP will be $\pm R$ by substitution in the normalised equation. If the coordinates of the vertices V_1, V_2, V_3 of the three BPs as shown in Figure 7.2 are substituted in the left hand side of the normalised equation of the PP, they will give distances from plane 123 *of the same sign* as given by substituting the coordinates of point T.

This latter test indicates that the bordering planes are 'upstanding' and may possibly be intersected or 'gouged' by the tool touching PP. This is the meaning of *interference*.

If the substitution gave a signed distance of opposite sign to that given by T, the vertex would be on the opposite side of the PP or 'downstanding' and interference could not occur.

Assuming the situation depicted in Figure 7.2, representing a potentially interfering situation, the equations of each facet can easily be found and normalised and distances u_1, u_2, u_3 computed.

The points at which the projectors u_1, u_2, u_3 intersect their respective facets may be inside or outside the triangular perimeter. If outside by a distance greater than $v = \sqrt{(R^2 - u^2)}$ the sphere will not gouge that facet; if closer to the triangle than v or actually inside, interference will occur. A treatment to resolve that point is rather tedious but is carried out by NEWERSUE by computations whose algorithms are outlined fully in the original references quoted above. The details will not be given here.

Once the normal distances of all facets within reach of the tool from the position T have been found, the tool is finally raised vertically (in direction z)

Figure 7.2 Tool-facet geometrical relationship with upstanding bordering facets interfering with the tool.

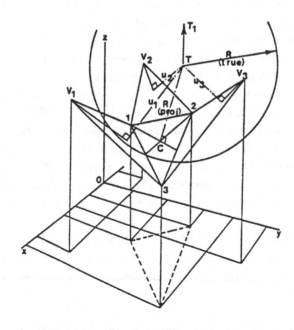

in computation to position T′ at which it just touches that facet with which it formerly interfered most severely.

The raising of the tool from T to T′ means that facet PP and some others will not have been touched or 'pointed' so that part of a complex surface will not have been well shaped. Thus, whenever the tool has been withdrawn, the program NEWERSUE writes another, and then another set of tool positions for a 'cascade' of tools, each half the diameter of the last, and arranges subsequent visits by such tools to the places missed by foregoing larger tools. This process of sending in ever smaller tools to locations of high localised curvatures is the means by which arbitrary surfaces of highly variable curvature, as encountered in human anatomy and terrestrial landscapes, can be accurately machined, starting with heavy removal of unwanted raw material with large tools.

4. Modes of Progress

In the dynamics of machining, the tool must move somehow from one corrected point T′ to another and in some systematic way must visit all points T′ for all facets of the polyhedron. At some point in the computations, each and every facet and each of its neighbours is examined for interference and points T′ found. In doing this the program NEWERSUE must progress in some orderly manner from one facet to another and the sequence of tool positions T′ to be visited by the tool is stored in readiness for actual machining.

The manner in which both the computations and the tool progress from one facet to the next is called the mode of progress.

The key to this mode of progress is the topologically rectangular nature of the polyhedron – this is fundamental. Three dimensional data (x, y, z) can be recalled by reference to row and column intersecting at nodes. Facets may be formed by joining (conceptually) either one of two sets of diagonals of the (curvilinear) quadrilaterals shown above in Figures 1.10 and 1.11. Then centroids of the resulting triangular planes may be visited for the tool-positioning and interference computations in one of the four modes shown in Figure 1.11 .

Two of these modes are 'zig zag'. Two others tend to be 'linear', passing twice, up and down, between each pair of rows. The latter give a relatively fine groove-like machining similar to the mode widely used with parametric surfaces. The former 'zig zag' tracking gives a distribution of asperities rather than tracks.

Any of these modes may be chosen: the logic of progress between facets is easily varied.

If progress of the tool centre is vectorial (linear) from T'_i to T'_{i+1} it will 'gouge' through the diagonal separating facets. This constitutes another minute error in this general approximate theory. In the earliest version of NEWERSUE called SUMAIR which was executed on a three axis machine with two controllers, one time-shared between y and z, the following logic, which incidentally avoided this 'diagonal gouging', was applied of necessity. If the next T' was higher than the last, the z increment was applied first followed by the x and y. If the next T' was lower than the last, the x and y motions were first applied and then the z. Even though modern machines move vectorially as a rule, this two step approach may yet have some value.

5. NEWERSUE Expressed Vectorially

Referring to Figure 7.3, the vector equation of a plane passing through points P,Q,R which are not collinear is:

$$\mathbf{x} = \mathbf{p} + \lambda (\mathbf{q} - \mathbf{p}) + \mu (\mathbf{r} - \mathbf{p}); \quad \lambda, \mu \text{ are scalar.}$$

Let **u** be the unit normal vector to the plane PQR. Then since $(\mathbf{q} - \mathbf{p})$ and $(\mathbf{r} - \mathbf{p})$ are each normal to **u**,

$$\mathbf{x} \cdot \mathbf{u} = \mathbf{p} \cdot \mathbf{u} = k; \quad k = \text{scalar.}$$

The unit vector, **u** is

Figure 7.3 Graphics of the vector representation of POLYHEDRAL NC[®] theory.

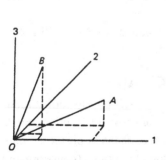
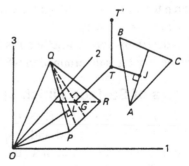

$$(((q - p) \wedge (r - p)) / |(q - p) \wedge (r - p)|) = u$$

The centroid G is given by:

$$g = {}^1\!/_3(p + q + r)$$

The tool-centre position T at distance ρ from G is:

$$t = {}^1\!/_3(p + q + r) + \rho u$$

Let the normal unit vector of any other neighbouring plane ABC within reach of the tool be vector v. Then the equation of the normal to the plane ABC from T is:

$$x = t + \upsilon v \quad (\upsilon \neq 0)$$

The distance of the plane ABC from the origin is l such that $x.v = l = a.v = b.v = c.v$, where x defines any point in plane ABC. Then the point of intersection, J, of the normal to plane ABC from point T with that plane is:

$$j = t + \upsilon v$$

where $\upsilon = l - t.v$ = distance from T to J.

Let the unit vector in direction 3 be e_3. If the tool radius is ρ (the distance from G to T) and υ is less than ρ, the tool will intersect the plane ABC in a small circle (of a sphere) and thus interfere with it.

To avoid such interference, the tool centre is displaced from T to T' a distance d in positive direction 3 until the distance from T' to plane ABC is less than ρ for every facet within reach of the tool at the x, y position of T (also of T'). Then $t' = t + d\, e_3$.

The determination of whether J is inside or outside of ABC will not be pursued here but it is determined in NEWERSUE.

6. Errors of Surface Finish in NEWERSUE

In the latest version of NEWERSUE the material left by the tool directly above each node is computed and graphic plots of error distributions made available. For each tool position T' the circular tool-shank is projected onto the $x\, y$ coordinate plane. A vertical line (direction z) is drawn (conceptually) through every grid-node within that circle to intersect the tool-sphere at a cal-

culated height z_i. This height will in general be different from the given surface-point. As the tool progresses over the field, the difference between the
latest calculated height z_i and the given point is repeatedly upgraded in a file
recording all grid-points. In the end, a perspective plot of residual errors –
heights of material not removed above surface-points – can be produced.

7. Program SUPERSUE

Whereas NEWERSUE is designed to place a spherical tool tangentially to a
plane facet determined by 3 polyhedral vertices, SUPERSUE aims to keep
those vertices exterior to the volume contained by its spherical cutting-end.
SUPERSUE thus has the character of *point avoidance* rather than *plane
avoidance*. These two concepts are illustrated and compared in Figure 7.4.

SUPERSUE suits well the orthogonal and topologically rectangular grid
used by the interpolating and graphic plotting routine TRUEPERS. Rows
and columns in the x, y coordinate plane are orthogonal straight lines of constant separation or pitch, Δx and Δy.

In SUPERSUE the tool axis is, as usual, maintained parallel to the coordinate axis of z and in machining is moved to and placed successively at each
node in every row. The spherical tool-tip is placed at a computed height
directly above each node. The increments of x along rows and the increment
of shift to a next row in direction y are constant: there is thus no offsetting
computation as with NEWERSUE.

Figure 7.4 NEWERSUE and SUPERSUE compared graphically.

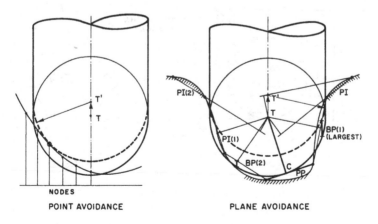

For a start the tip of the tool-sphere is imagined to touch the surface point whose coordinates are stored at projected position x_{ij}, y_{ij} in the typical ith row and jth column as in Figure 1.12. Next, all nodes within the projected circle of the tool shank are identified. Then a calculation is carried out for each such point to discover where the tool centre T$'$ should be at axis location x_{ij}, y_{ij} so that the tool sphere just passes through the (different) point tested. Then initial position T at $z = z_{ij} + R$, where R = tool radius, is adjusted to T$'$. If a previous point had been tested and an adjustment made, the latest highest value of T$'$ is retained at storage location x_{ij}, y_{ij}. In this way a highest point T$'$ is retained and stored in a file providing for all nodal positions. The scanning scheme for this computation is shown in Figure 7.5.

The processes just described can also be understood from the two dimensional analogue sketch in Figure 1.12 showing why the tool should be moved from the full to the dashed position to avoid the shaded undercutting. For nodes just inside or just outside of the 'equator' of the sphere, there is a possibility of undercutting associated facets if the surface there is very steep. If this were to be corrected, an analysis similar to that of NEWERSUE would be necessary and this has not been done here.

So a file of tool centre positions T$'$ can be repeatedly updated as the tool-axis is placed at successive grid-points.

Figure 7.5 The scanning scheme of SUPERSUE.

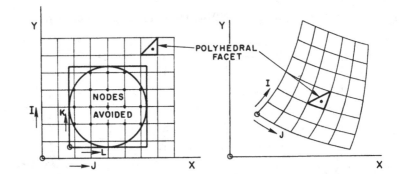

SUPERSUE   NEWERSUE

Another file of cut surface-points at each node x_{ij}, y_{ij} is also computed for all nodes within reach of the tool at a current tool-axis nodal location. A vertical line is drawn (conceptually) through each node within reach from T_{ij} to intersect the tool-sphere at that location. Thus the cut point in the surface is found for each node. As the tool moves from one node to another this file of cut surface-points at nodal locations is updated to the lowest point so far cut there. After a complete traverse of the whole field, these ultimate cut points can be compared with the initially given surface heights at each nodal location. The differences representing errors of cutting due to the approximate nature of the whole process may then be plotted perspectively to reveal the effect of those approximations.

The error plot of SUPERSUE may also be compared by subtraction with that for NEWERSUE using identical given surface data for the two independent treatments.

The determination of T'_{ij} and the resulting cutting errors are computed by four nested program loops, two to find a node for tool location and two to find all nodes within the projected circle of the tool-shank.

The initial tool centre positions are calculated using a large tool to remove the bulk of the material. The error due to the inability to machine the surface-details with a large tool is calculated. The program SUPERSUE automatically calculates the new tool centre position for a tool one-half the diameter of the previous tool to visit only those surface-points whose error is greater than the required accuracy. With a smaller tool, finer surface detail can be machined. Successively smaller tools are used to machine the areas of fine detail until the program reaches a minimum tool diameter specified by the user.

One advantage of SUPERSUE is that all x and y motions are by known and fixed increments, Δx and Δy. Thus these movements only need counters: there is no computation of offsets leading to the location of T' at points other than nodes.

Another advantage is that several work-heads can operate simultaneously on the same work table of a machine. Only the setting height z_{ij} of each tool differs in each head, assuming that each head is cutting completely different surfaces. Two controllers are needed to apply the increments of x and y (and these need only have indexing capacity) and each head requires a dedicated controller to handle z_{ij}.

8. Comparison between NEWERSUE and SUPERSUE

Although the result of NEWERSUE and SUPERSUE is similar (to find the cutter location data (CLD) for a ball end mill without interference), the application for each method is quite distinct, depending on the data, the size of the data, and the computational and machining time. These differences are due to the fact, illustrated in Figure 7.4, that NEWERSUE is a plane avoidance algorithm and SUPERSUE is a point avoidance algorithm.

SUPERSUE requires the data to be orthogonally and topologically rectangular. Such data can be obtained from programs such as TRUEPERS. The principle behind SUPERSUE is that the tool radius should be larger than the grid size. Otherwise, interference cannot be checked. Preferably, the tool radius is several times larger than the grid spacing.

NEWERSUE, on the other hand, requires the data to be *topologically* rectangular (same number of points per line for a given set of lines). Therefore, data from surfaces like Bézier and B-spline as well as data from TRUEPERS can be used. The spacing between points may be comparable to the tool radius.

The computational time (time to calculate the CLD) is much shorter for SUPERSUE than for NEWERSUE for several reasons. First, NEWERSUE processes almost twice $(2 \times (N-1) \times (M-1)$ to be exact) the number of CLD for a given set of N by M points whereas SUPERSUE calculates exactly N by M numbers of CLD. Second, NEWERSUE calculates the centroid, coefficients and normal for each plane before interference checking begins. Third, plane interference checking is more computational intensive than point interference checking.

The size of data that can be used in NEWERSUE is smaller compared to SUPERSUE because NEWERSUE stores the coefficients and centroids of each plane inside the memory. SUPERSUE does not. For a given size of memory in the computer, more data are stored in NEWERSUE, resulting in a smaller grid size data. Other methods of storing less data were investigated but the computational and disk access time increased.

There are two basic modes of machining in POLYHEDRAL NC®: 'point-to-point' (moving vectorially in 3 dimensions) and 'true pointing' (moving in 2 dimensions and lowering the tool to the required depth). In both cases, the initial machining mode should be moving vectorially with a large tool to remove the bulk of the material. On the second pass, either mode can be used.

As the authors have experienced, 'true pointing' may be slower than 'point-to-point' because 'pointing' requires the tool to move up to clear the workpiece, then to a new position and finally down to the new depth. In some cases, re-machining the entire surface with a smaller tool may be quicker than re-machining only those points or facets where interference has occurred with a larger tool. Of course, the computational time to calculate the CLD must be considered in determining the shortest time to produce a given surface. POLYHEDRAL NC® allows the user to decide, from these factors, the optimal machining mode.

9. Program CURVATUR

Using the coordinated surface-point data produced by TRUEPERS or data obtained by modern optical measuring systems arranged in a similar array, the maximum principal curvature at nodes of a smooth surface represented approximately by the data may be found by using finite difference (numerical) calculus.

The Gaussian expression for the two principal curvatures $K_{1,2}$ at a point on a surface is:

$$K_{1,2} = \text{mean} \pm \sqrt{((\text{mean})^2 - \text{Gauss})}$$

where 'mean' is the *mean curvature* and 'Gauss' is the *Gaussian curvature*, fundamental entities in differential geometry.

The numerical evaluation of these fundamental quantities is given in the documentation of Program CURVATUR.

The *reciprocal* of the greatest value of principal curvature K_1 gives the smallest local radius of curvature. Spherically-ended cutting tools of any larger radius than this, if caused to touch a local facet, would interfere with neighbouring facets unless withdrawn as done by NEWERSUE or SUPERSUE. So if it is desired to cut a complete surface patch with a single tool in one pass, the curvature computation indicates the maximum size of tool that can be used.

If, for some reason, the largest tool to be used has been arbitrarily determined, the maximum curvature it can handle is found as the reciprocal of its radius. Any surface curvature larger than this marks out a region of the surface where there is a potential for interference; where withdrawal of the larger tools of a 'cascade' is necessary and where smaller tools must visit later to cut the surface more accurately.

Perspective plots of local maximum principal curvature of a surface defined numerically can be plotted through a second application of TRUEPERS, taking curvature instead of coordinate z as ordinate. Also, any curvature greater than a nominated value may be plotted to reveal graphically where the potentially interfering regions of a surface patch are.

Program CURVATUR is useful in providing guiding information in the use of both NEWERSUE and SUPERSUE; it has thus been kept as an independent program.

10. Converting Manual to CNC Milling Machines

A new CNC milling machine can be expensive. Retrofitting manual milling machines to CNC milling machines can now be done economically. The milling machine should be of a good quality with recirculating ball and lead screw. Many companies sell DC servo motors and encoders which can be mounted on each of the axes of the machines. The encoders allow the milling machine to have a closed loop or feedback system to indicate to the controller the position of the table and tool. Some early CNC machines had stepping motors to determine the position of the table and tool. In addition, commercially available hardware/software systems can transform a personal computer to a milling machine controller. The personal computer can first be used to generate surfaces and tool positions. Then by running software supplied by the hardware/software system, the computer is converted to a controller console. Some systems allow for backlash in the machine which may be present in some milling machines. In most cases, the signal from the computer to drive the servo motors is not strong enough; therefore, power amplifiers are required to obtain a signal with enough current to drive the DC servo motors. With such an economical retrofit, one can produce models as described in this book.

Chapter 8

APPLICATIONS

1. Introduction

Since it was first devised in response to a challenge to define and machine complex surface-shapes in engineering, POLYHEDRAL NC® has been applied to the replication of many other items in both engineering and medicine. Most of these applications are described in detail in publications, in the book *Sculptured Surfaces in Engineering and Medicine* (Duncan and Mair, 1983), and in the references quoted there.

When first applied, a mainframe computer was used to store and process the data of this data-storage system. Following developments in micro- and mini-computers during the last few years, the programs of POLYHEDRAL NC® are manageable on IBM PC compatible computers and data storage is no longer an obstacle or a disadvantage. In fact, the use of data storage enables such problems as interference, filleting and non-invasion to be readily and more conveniently solved, even for analytical surfaces.

The documented programs appended to this book are updated versions of the most general and significant of those which have been devised and used over some 15 years with new additions.

In this chapter, applications in various categories are described briefly with reference to sources of more detail.

2. Analytical Surfaces

If a surface has a Cartesian or parametric equation a computer can calculate surface-points or, with little more complication, normals to that surface and hence CLD.

Many engineering surfaces are compounds of analytical pieces intersecting discontinuously at space-curves of interpenetration. This is common in diemaking. Finding those curves of interpenetration analytically is often no simple matter, nor is the determination of CLD at such junctions. However, the combination of the *Method of Highest Point*, executed through the program GEN7, with the anti-interference features of NEWERSUE or SUPERSUE to control machining, automatically determines such junctions and fillets them with a spherical radius.

Items of such a nature which are illustrated in *Sculptured Surfaces in Engineering and Medicine* are:

- TRIUMF meson accelerator inflector (Figure 8.1)
- Pipe tee
- Piston core
- Automobile rear lamp reflector punch
- Vacuum cleaner moulding
- Various plastic bottle moulds
- Automotive headlamp reflector die

3. Analytical Surface-scaling

As explained in connection with proportional development, a surface defined within a unit square by $z = f(x, y)$ can be stretched to 'fit into' an arbitrary outline. If the points defining the surface are nodes of a conventional parametric grid, they remain topologically rectangular after stretching and the POLYHEDRAL NC® approach may be applied to their machining.

Three items in which an analytical surface was stretched like this are:

- Mould for a plastically moulded sole for a lady's shoe
- Mould for a cleated military boot
- Top-plate for a violin (Figures 2.13 and 8.2)

4. Proportionally Developed Surfaces

In cases where edge-slopes are not constrained by slope-requirements or when global surfaces are composed of symmetrical pieces, the computer-aided version of a former general graphical method – Proportional Development – may be applicable.

If adjustment within a global boundary is required the superposition of a bi-beta function often suffices.

The book, *Sculptured Surfaces in Engineering and Medicine*, gives many examples among which are:

- Automobile roof-panel die
- Aircraft engine nacelle
- Ship's hull with adjustment
- Crawler side bar (Figure 8.3)
- Plastic shoe-heel moulding

Figure 8.1 The TRIUMF meson accelerator inflector.

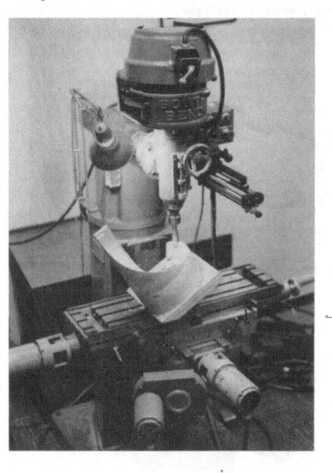
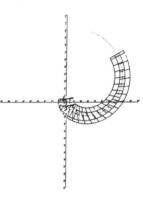
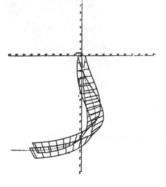
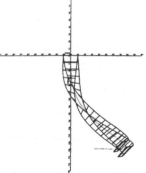

Figure 8.2 Top plate of violin with tool path plot.

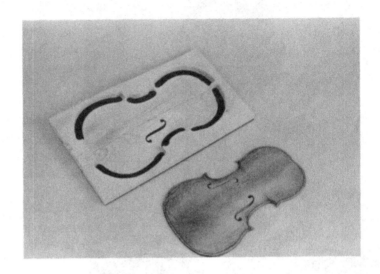

Figure 8.3 Crawler side bar (fillets).

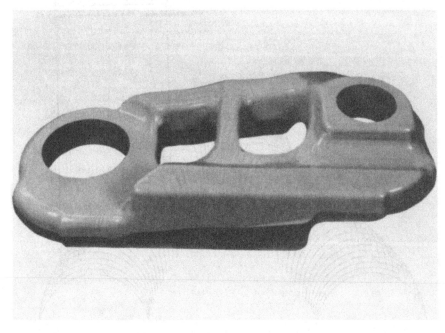

5. Duct Surfaces

Many surfaces are tubular in nature, the cross-section being normal to a 'spine' – a twisted space-curve – at all points along it. Sometimes such a surface has to be moulded as a thin shell, literally a tube, and, perhaps, in two half-sections joined along a parting surface.

Thinking of such surfaces as long cavities, the parameters are, as always, two: s, the arc length along the spine and θ, the angle of a vector in the normal plane to points on the perimeter of that section. The radius of that vector may be a function of parameter s.

Figure 8.4 Bifurcated equi-area carburetta duct defined by super ellipses of variable parameters.

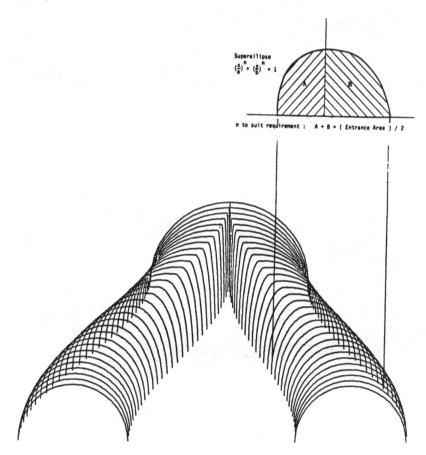

Surfaces defined by points found by incrementing s and θ may in turn be defined by Cartesian coordinates in a topologically rectangular array and the surface so defined is at once ready for machining by POLYHEDRAL NC®.

Some surfaces which have been so machined are:

- Jug mould
- Carburetta bifurcated duct (equi-area duct) (Figure 8.4)
- Klein bottle
- Pipe bend transformer – round to square with equal area
- Engine manifold core box (Figure 8.5)

6. Human Anatomy Replication

Data for human anatomy are often measured by one of the techniques described earlier. Frequently these data are processed by program TRUEPERS or otherwise filtered though they are often used as 'raw data'. The pointing process implies that some hand-finishing and thus smoothing may be applied manually.

The optical scanner of Boulanger et al. collects data directly in a topologically rectangular array; so no interpolation by TRUEPERS is essential though the processing of such data smooths it by local interpolation.

Photogrammetry, shadow-moiré, peri-contourography, silhouetting and the method of cylindrical envelopes have all been used by the authors for data collection followed by polyhedral machining.

In conjunction with CEMAX Inc. several skull replicas were machined from data obtained by Computed Axial Tomography (CAT) using a proprietory version of NEWERSUE.

Medical items so machined include:

- Amputee stump socket mould
- Human knee
- Cosmetic cover mould for lady amputee
- Various foot lasts
- Various human faces
- Ancient Japanese skull
- Tooth replica
- Bust of a girl (Figure 8.6)
- Human skull bone (from CAT scan of a cadaver) (Figure 3.6)

7. Landscapes and Seabeds

Several of these have been produced from digitised points on survey maps or from soundings at random positions on a marine chart, in both cases

Figure 8.5 Engine manifold core box defined and machined by POLYHEDRAL NC®. (Beijing University of Aeronautics and Astronautics. By courtesy of Professor Tang Rongxi.)

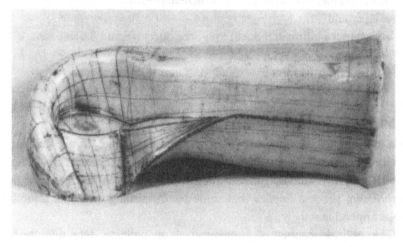

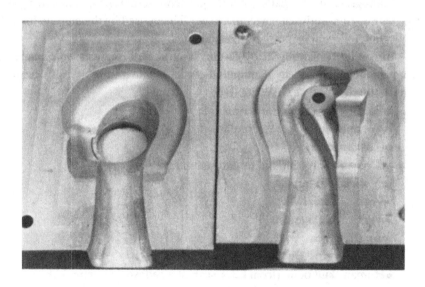

processed by TRUEPERS to generate and display surface-points at nodes of a rectangular grid.

These items are:
- Georgia Strait seabed model
- Point Atkinson model (Figure 8.7)
- Lions Bay model

Figure 8.6 Machining by scrolling about a spindle of the bust of a girl. (Data obtained by high speed scanning by Hymarc Engineering and NRC, Ottawa.)

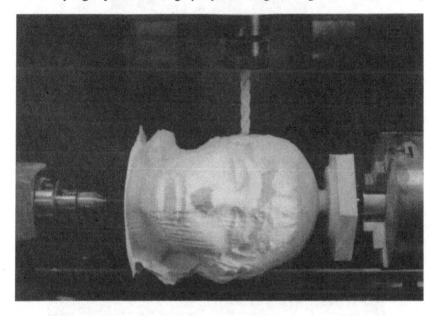

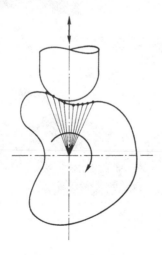

Figure 8.7 Terraced model of terrain at Point Atkinson, B.C., from aerial survey.

Figure 8.8 Smooth model of a dam site machined by NEWERSUE from aerial survey.

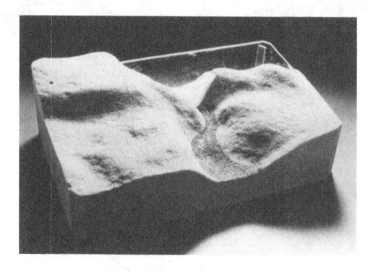

- B.C. Hydro dam site model (Figure 8.8)
- Isle of Anglesey model
- U.B.C. satellite survey model

8. Turbine, Fan and Propeller Blades

Various blades have been defined and machined:
- Fan blades (student exercise)
- Joukowski airfoils
- Marine propeller model (by Dr. G. W.Vickers)

9. Tetrahedral Applications

These have been in two contexts, one to find all the mechanical properties of arbitrarily shaped bodies and the other to check for interference in robotics. Examples of the first application are given with the documentation herewith. A brief account of the second application is found in the proceedings of the International Conference on Computer-aided Production Engineering (Duncan, 1986).

Appendix A

SUPPLY OF PROGRAMS

The programs documented and explained in this work can be obtained in executable form on five IBM PC compatible disks.

The DEMONSTRATIONS enabling potential users of the general programs to assess the POLYHEDRAL NC® approach are available separately on two disks at a nominal cost of the disk, postage and handling.

Assistance with initial use of the general programs is available.

To obtain these disks, write to or call:

2184 E. 35th Ave.
Vancouver, BC
CANADA
V5P 1B9

Tel (604) 327-4564
(604) 228-2781
(fax) (604) 327-4564

DEMONSTRATION INSTRUCTION I

Introduction:

POLYHEDRAL NC® is a set of programs which enables the user to define surfaces for machining on numerical control milling machines. Figure A1 shows the relationships between each program and its input and output files. (The blocks represent the program names and the double boxes, which symbolise the old fashion 'stack of cards', represent data files for input/output.)

The data used in this demonstration are obtained from a set of boundary curves in plan and elevation views (see Figure A2). The equations of these curves were used as input into PROPDEV to generate a set of surface points along twisted space curves. The surface points, as shown in Figure A3, are determined by fractions of range, alpha and beta, of any two of the three coordinates along the boundaries (which is similar to the u, v parametric character of a Bézier surface).

Purpose:

To generate a file of Cutter Location Data (CLD) points using program NEWERSUE, followed by a file of RS-274D standard controller codes using program GCODE, using a curvilinear topologically rectangular array of surface points as input.

Installation:

This section is for installing the demonstration programs onto the hard disks. This procedure is required to be performed once. The installation will create a subdirectory called 'DEMO1'.

    1. Insert the demonstration disk into drive A:

    2. Set the default drive to A:

    3. Install the programs onto drive C: by typing:

    insdemo1 c:

Procedure:

Before starting the demonstration, change the default directory of the hard disk to DEMO1 by typing:

Figure A1 Flow chart for POLYHEDRAL NC® system.

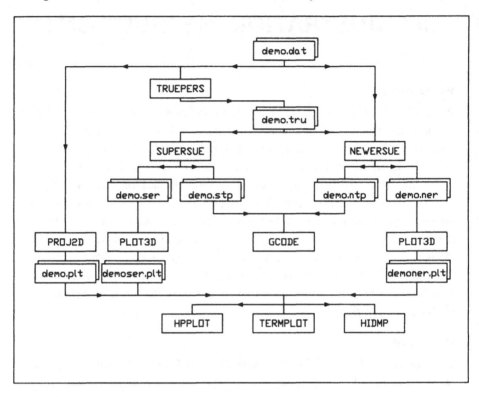

Figure A2 Plan and elevation view of demonstration surface.

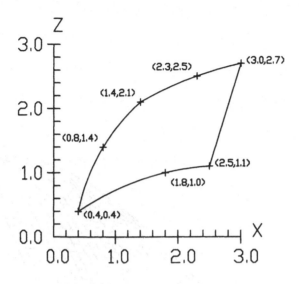

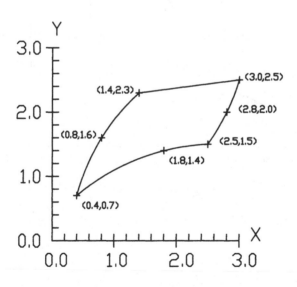

Figure A3 Perspective view of demonstration surface.

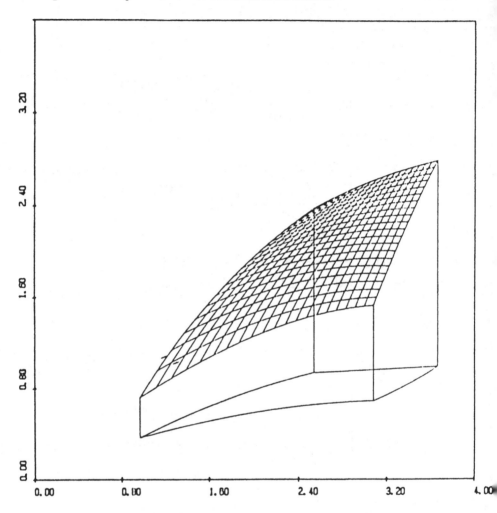

cd \demo1
PROJ2D
to generate a plot file to view the surface. The plot file, a file containing the plotting instructions, is called 'demo.plt'. The following questions are asked by the program. The user should respond as suggested below.

> Enter name of input file: demo.dat
> Enter NX, NY: 21,21
> Enter Z scaling factor: 1.0
> Enter name of output file: demo.plt
> Enter eye position: 5,-5,3
> Reading in data
> Projecting data
> Stop - Program terminated.

PMAXMIN demo.plt
to determine the window of the plot file (the range in which the picture exists).

TERMPLOT demo.plt
to view the plot file on the screen. Figure A3 shows the result of the surface.

NEWERSUE
to generate the CLD for a set of curvilinear data points using NEWERSUE with anti-interference. The CLD is stored in the file 'demo.ntp'. The following questions are asked by the program. The user should respond as suggested below.

> Enter input file name: demo.dat
> Enter output file name: demo.ntp
> Enter origin (3F10.3) 0.,0.,0.
> Enter input scaling factor FX,FY,FZ: 1.,1.,1.
> Enter RAD: 0.25
> Enter output scaling factor OFX,OFY,OFZ: 1.,1.,1.
> FX = 1.0000 FY = 1.0000 FZ = 1.0000
> RAD = .2500
> FX = 1.0000 OFY = 1.0000 OFZ = 1.0000

Check interference (1 = not check, 0 = check): 0
Enter number of pass (1 or 2): 1
Enter calculation tolerance (0 means .001): 0
Enter smallest tool used (smallest = .09375): 0
Enter print switch (0 = off,1 = on): 1
Enter data output: (1 = tool position,2 = centroids): 1
Enter direction of triangle (0 or 1 = /, 2 = \): 1
NO. OF PASSES = 1 TOLERANCE = .001 MINIMUM RADIUS = .094
Enter data size in X and Y: 21,21
Enter part of X to machine, (IM1,IM2) (IM1 = 0 means all): 0,1
Enter part of Y to machine, (JM1,JM2) (JM1 = 0 means all): 0,1
NO. OF ROWS 21   NO. OF COLUMNS 21
  MACHINING ROWS 1 21
  MACHINING COLUMNS 1 21
Error analysis (1 = yes,0 = no): 1
NO. OF PLANES =   800
INTERFERENCE OCCURRED AT   0 PLANES
RANGE OF X = 2.9469 RANGE OF Y = 2.2939 RANGE OF Z = 2.5927
MINIMUM X = .0000 DISTANCE FROM ORIGIN = .0000
MAXIMUM X = 2.9469 DISTANCE FROM ORIGIN = 2.9469
MINIMUM Y = .0000 DISTANCE FROM ORIGIN = .0000
MAXIMUM Y = 2.2939 DISTANCE FROM ORIGIN = 2.2939
MINIMUM Z = .2500 DISTANCE FROM ORIGIN = -2.5927
MAXIMUM Z = 2.8427 DISTANCE FROM ORIGIN = .0000
  ORIGIN IS:
    .0000   .0000   2.8427
  writing data to file
Total number of node points machined is   441
Average error at each node is   .007901
Data are stored in file demo.ntp
Screen display is stored in file TOOLDAT.DAT
If analyse error, errors stored in file ZDIFF.DAT
Stop - Program terminated.

GCODE

to generate a machining code for the NC machine based on the EIA RS-274D standard. Different controllers use different standards. This standard is used by many FANUC controllers. The data file which stored the G-code is called 'demo.nc'. The following questions are asked by the program. The user should respond as suggested below.

> Enter units? (1 = inch, 0 = mm) 1
> Enter feedrate? (real) 35.
> Enter spindle speed? (int) 1500
> Enter spindle direction? (1 = cw, 2 = ccw) 1
> Enter nc program number 123
> Enter starting tool position form origin? -2.,0.,0.
> Enter size of material? (X,Y,Z) 6.,5.,4.
> Enter name of input file? demo.ntp
> Enter name of output file? demon.nc
> Stop - Program terminated.

DEMONSTRATION INSTRUCTION II

Purpose:

To transform random data (topologically rectangular data in a curvilinear array can be considered as random data) to geometrically rectangular grid data using program TRUEPERS and then generate Cutter Location Data files using programs SUPERSUE and NEWERSUE followed by files of RS-274D standard controller codes from both methods using program GCODE. Plotting with hidden line removal of the transformed data will also be done by PLOT3D.

Installation:

This section is for installing the demonstration programs onto the hard disks. This procedure is required to be performed once. The installation will create a subdirectory called 'DEMO2'.

    1. Insert the demonstration disk into drive A:

    2. Set the default drive to A:

    3. Install the programs onto drive C: by typing:

      insdemo2 c:

Procedure:

Before starting the demonstration, change the default directory of the hard disk to DEMO2 by typing:

cd \demo2

TRUEPER1

to transform the curvilinear data to a geometrically rectangular grid. The transformed data are stored in a file called 'demo.tru'. To obtain a smooth surface, the user should process the data twice. For the first pass, the following questions are asked by the program. The user should respond as suggested below.

106

Program TRUEPERS: locally interpolates a file of random points to a topologically rectangular grid.

Enter input file name: demo.dat

Enter output file name: demo.tru

Enter number of points to be interpolated in X: 27

Enter xmin, xmax: .4,3.0

Enter number of points to be interpolated in Y: 19

Enter ymin, ymax: .7,2.5

Enter CAY, NRNG, NSM (real,int,int): 20.,8,16

Blanking data? (1 = yes) 1

Enter name of blanking data file: demo.bnk
    441 data points read in
     6 blanking points read in
SUBROUTINE ZGRID
10 W = 1.020423 ROOT = .355746 DZMAX/ZRANGE = .0000029
SUBROUTINE SMOOTH
Stop - Program terminated.

TRUEPER2

reprocess the data for the second pass. The following questions are asked by the program. The user should respond as suggested below. The final data are stored in file 'demo2.tru'.

Program TRUEPERS: locally interpolates a file of random points to a topologically rectangular grid.

Enter input file name: demo.tru

Enter output file name: demo2.tru

Enter number of points to be interpolated in X: 53

Enter xmin, xmax: .4,3.0

Enter number of points to be interpolated in Y: 37

Enter ymin, ymax: .7,2.5

Enter CAY, NRNG, NSM (real,int,int): 50.,18,14

Blanking data? (1 = yes) 1

Enter name of blanking data file: demo.bnk
    513 data points read in
     6 blanking points read in
SUBROUTINE ZGRID

10 W = 1.309380 ROOT = .790646 DZMAX/ZRANGE = .0016298
SUBROUTINE SMOOTH
Stop - Program terminated.

The user now has two options: to view or to generate the CLD for the surface. The demonstration will illustrate the plotting first. Before viewing the surface, a plot file needs to be generated.

PLOT3D

to generate the plot file with hidden line removal. The plot file generated will be called 'demo2.plt'. The following questions are asked by the program. The user should respond as suggested below.

Program PLOT3D: plots a 3D surface with hidden line removal
Enter the name of input file: demo2.tru
Enter the name of plot file: demo2.plt
Enter the number of points in X direction: 53
Enter the number of points in Y direction: 37
Enter values for ZBASE, ZMAG, R, THETA, PHI: -.05, .5, 7.6811, -45., 23.
Enter every nth point to plot: 1
Enter paper size: Xmin, Xmax: 0.,10.
                  Ymin, Ymax: 0.,10.
Enter contour level: (0 = none) 0
    .0500   .0500
   3.0000   .4000
   2.5000   .7000
SUBROUTINE XLINES
SUBROUTINE YLINES
SUBROUTINE BLINES
Stop - Program terminated.

TERMPLOT demo2.plt

to view the plot file on the screen. The window for the plot is the paper size specified by program PLOT3D. Program PMAXMIN is not necessary. Figure A4 shows the result of the plot.

NEWERSUE

to generate the CLD for a set of topologically rectangular data points using NEWERSUE with anti-interference. The CLD is stored in the file demo2.ntp'. The following questions are asked by the program. The user should respond as suggested below.

Enter input file name: demo2.tru
Enter output file name: demo2.ntp
Enter origin (3F10.3) 0.,0.,0.
Enter input scaling factor FX,FY,FZ: 1.,1.,1.
Enter RAD: 0.25
Enter output scaling factor OFX,OFY,OFZ: 1.,1.,1.
FX =   1.0000   FY =   1.0000   FZ =   1.0000
RAD =   .2500
FX =   1.0000   OFY =   1.0000   OFZ =   1.0000
Check interference (1 = not check, 0 = check): 0
Enter number of pass (1 or 2): 1
Enter calculation tolerance (0 means .001): 0
Enter smallest tool used (smallest = .09375): 0
Enter print switch (0 = off,1 = on): 0
Enter data output: (1 = tool position,2 = centroids): 1
Enter grid data: (0 = grid,1 = not grid): 0
Enter direction of triangle (0 or 1 = /, 2 = \): 0
NO. OF PASSES = 1 TOLERANCE = .001 MINIMUM RADIUS = .094
Enter data size in X and Y: 53,37
Enter part of X to machine, (IM1,IM2) (IM1 = 0 means all): 0,1
Enter part of Y to machine, (JM1,JM2) (JM1 = 0 means all): 0,1
NO. OF ROWS 53   NO. OF COLUMNS 37
  MACHINING ROWS 1 53
  MACHINING COLUMNS 1 37
NO. OF PLANES = 3744
INTERFERENCE OCCURRED AT 0 PLANES
RANGE OF X = 3.2333 RANGE OF Y = 2.7333 RANGE OF Z = 2.8355
MINIMUM X = .0000 DISTANCE FROM ORIGIN = .0000
MAXIMUM X = 3.2333 DISTANCE FROM ORIGIN = 3.2333
MINIMUM Y = .0000 DISTANCE FROM ORIGIN = .0000
MAXIMUM Y = 2.7333 DISTANCE FROM ORIGIN = 2.7333
MINIMUM Z = .0000 DISTANCE FROM ORIGIN = -2.8355

Figure A4 Demonstration surface smoothed twice.

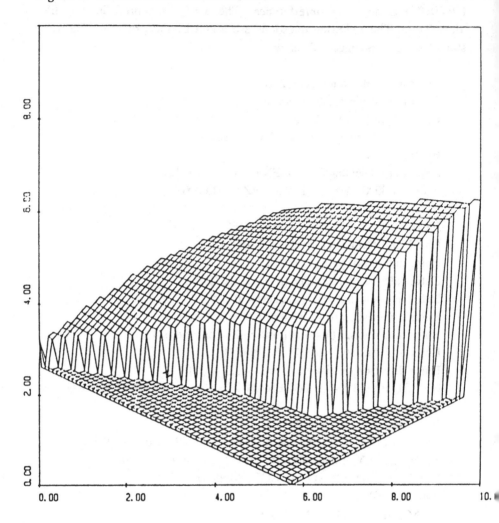

MAXIMUM Z = 2.8355 DISTANCE FROM ORIGIN = .0000
ORIGIN IS:
  .0000   .0000   .2500
writing data to file
Data are stored in file demo2.ntp
Screen display is stored in file TOOLDAT.DAT
If analyse error, errors stored in file ZDIFF.DAT
Stop - Program terminated.

GCODE

to generate a machining code for a NC machine based on the EIA RS-274D standard. Different controllers use different standards. This standard is used by many FANUC controllers. The data file which stored the G-code is called 'demo2n.nc'. The following questions are asked by the program. The user should respond as suggested below.

Enter units? (1 = inch, 0 = mm) 1
Enter feedrate? (real) 35.
Enter spindle speed? (int) 1500
Enter spindle direction? (1 = cw, 2 = ccw) 1
Enter nc program number 123
Enter starting tool position from origin? -2.,0.,0.
Enter size of material? (X,Y,Z) 6.,5.,4.
Enter name of input file? demo2.ntp
Enter data type: (1 = supersue, 2 = newersue) 2
Enter name of output file? demo2n.nc
Stop - Program terminated.

SUPERSUE

to generate the CLD for a set of topologically rectangular data points using SUPERSUE with anti-interference. SUPERSUE only applies to topologically rectangular data. The CLD is stored in the file 'demo2.stp'. The following questions are asked by the program. The user should respond as suggested below.

Enter input filename: demo2.tru
Enter output filename: demo2.stp

    Enter NX 53
    Enter NY 37
    Enter every Nth point to machine in X 1
    Enter every Nth point to machine in Y 1
    Enter tool radius 0.25
    Please wait for computer to read in data
    Number of points to check in X is 5
    Number of points to check in Y is 5
    Max. error is 2.650799 at position 3.0000 2.4500
    Stop - Program terminated.

GCODE

to generate a machining code based on the EIA RS-274D standard. The data file which stored the G-code is called 'demo2s.nc'. The following questions are asked by the program. The user should respond as suggested below.

    Enter units? (1 = inch, 0 = mm) 1
    Enter feedrate? (real) 30.
    Enter spindle speed? (int) 1500
    Enter spindle direction? (1 = cw, 2 = ccw) 1
    Enter nc program number 123
    Enter starting tool position from origin? -2.,0.,0.
    Enter size of material? (X,Y,Z) 6.,5.,4.
    Enter name of input file? demo2.stp
    Enter data type: (1 = supersue, 2 = newersue) 1
    Enter name of output file? demo2s.nc
    Stop - Program terminated.

Note:

HIDMP has not been thoroughly tested. It works with a series of Houston Instrument plotters.

Appendix B

Installing Instructions
to
POLYHEDRAL NC® Programs

Disclaimer: The programs comprising POLYHEDRAL NC® have been thoroughly tested and used over a period of years. However, no guarantee of their performance is expressed or implied except as they may be used with the supervision of the authors or those trained ith their implementation.

Warning: The POLYHEDRAL NC® programs are copy protected. Therefore, all the programs must be recalled before any back-up and reformatting can be done on the hard disk, otherwise, the hard-disk version of the programs will not be executable after reformatting.

Since the preparation of the documentation, new compilers have become available which will enable the programs to handle array sizes at least twice as large as indicated in the documentation.

Initial Installation (performed once only)

1. Create a subdirectory on the hard disk to store the POLYHEDRAL NC programs by the MKDIR command in DOS:
   MKDIR C:\POLY_NC
2. Change the current directory to the POLYHEDRAL NC directory using the CHDIR command in DOS:
   CHDIR C:\POLY_NC
3. Insert the disk containing the programs into drive A:
4. Set the default drive to A:
5. To install the programs, type the command

INSTALL A: C:

6. Change the default drive back to C:
7. Change the current directory to the root directory by the command: CD\
8. If the file 'AUTOEXEC.BAT' already exists, edit the file and insert the following line to create a path for the programs.
   PATH C:\POLY_NC
   If the file does not exist, create a file with the name 'AUTOEXEC.BAT' and insert the line
   PATH C:\POLY_NC
9. Type the following command:
   AUTOEXEC

Installation after Initial Install

After the initial installation, future installation is much easier.

1. Change the current directory to the POLYHEDRAL NC directory using the CHDIR command in DOS:
   CHDIR C:\POLY_NC
2. Insert the disk containing the programs into drive A:
3. Set the default drive to A:
4. To install the programs, type the command
   INSTALL A: C:
5. Change the default drive back to C:

Recalling Programs from Hard Disk

1. Change the current directory to POLY_NC by
   CD C:\POLY_NC
2. Insert the disk containing the programs to recall
3. Set default to drive to A:
4. To recall the programs, issue the command:
   RECALL C: A:

To recall more than one disk, repeat the four procedures above.

Note:

All the programs must be recalled before backing up and reformatting the hard disk. Otherwise, the POLYHEDRAL NC® programs will not execute after reformatting the hard disk.

   After recalling and reformatting of the hard disk, follow the initial installation to install POLYHEDRAL NC® programs back to the hard disk.

CALTOOLP

Purpose:

CALTOOLP calculates the tool path which is one tool radius away from a two dimensional contour. The contour is represented by a series of points lying on the contour. The program calculates the tool path for cutting on the inside or the outside of a given contour but has no anti-interference check.

Method:

The program uses a numerical technique to find the tool offset.

Hardware:

IBM PC compatible personal computer running under DOS operating system with 640K of RAM and a hard disk or high density floppy disk.

Input:

ENTER THE DATA FILE NAME:

Contours are defined by (X,Y) coordinates. The input file consists of three real numbers each separated by a comma and/or one or more spaces. The first number represents the X coordinate, the second the Y coordinate and the third the Z coordinate respectively.

ENTER OUTPUT FILE NAME:

Enter the file name to store the tool offset.

ENTER PLOT FILE NAME:

Enter the plot file to store plotting commands.

ENTER TOOL DIAMETER:

ENTER DIRECTION, (I = IN, O = OUT):

If the contour is digitised in a counter-clockwise direction, 'I' will generate the tool offset on the 'inside' of the contour and 'O' will generate the tool offset on the 'outside' of the contour.

ENTER NOL (NUMBER OF LOOPS):
NOL is the number of loops the tool will circle.

Output:

The tool offset is given in a similar format as the input file. In addition, the plot file is generated for viewing the tool path on the screen or on a plotter.

Example:

Enter the name of the data file: *caltoolp.dat*
Enter the name of the output file: *caltoolp.tp*
Enter the name of the plot file: *caltoolp.plt*
Enter the diameter of the cutter: *0.5*
Which side in a ccw direction? (I = in,O = out): *I*
Enter the number of loops: *2*

Note:

A plot file is generated for displaying the original contour and the tool offset.

Figure B1 Tool offset paths from contour.

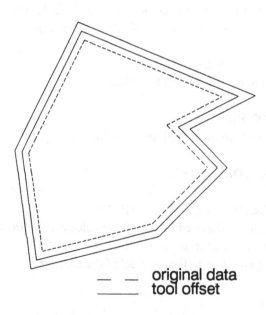

— — original data
———— tool offset

CURVFIT

Purpose:

To fit a piecewise, two dimensional second degree slope-continuous, plane curve through a set of digitised points with a chosen degree of 'tight' or 'loose' fit.

Method:

The 'pieces' of curve, which are strung together with slope continuity, all have the general second degree equation of the form:

$$Ax^2 + Bxy + Cy^2 + Dx + Ey + F = 0$$
or with $A \neq 0$;
$$x^2 + B'xy + C'y^2 + D'x + E'y + F' = 0$$

An arbitrary input curve is specified in terms of digitised points, randomly distributed along the curve as shown in the following figures. Certain special or control points such as end-points, maximum and minimum points, extreme points and points of inflexion are identified. At each of these points, a slope is determined *by judgement or selection* by means of a second off-curve point.

The pieces as defined above have 5 independent coefficients; 5 conditions are therefore required to determine the values of the coefficients. If one 'piece' of this form is to span an interval between the given control points passing through or near the more numerous digitised points, we may use the coordinates of the two end-points and a 'mid-count' point to give three equations and coefficients, and two end-slopes to give two more. Thus five conditions give five coefficients.

The mid-count point is then iterated (oscillated) in x and y independently and the root mean square (RMS) departure of the digitised and corresponding curve-points minimised by this adjustment.

If the resulting RMS departure is still greater than a stipulated value, the program CURVFIT automatically determines a *weighted mean slope* at the mid-count point and continues as before to find two pieces, tangential at the former mid-count point and adjusted by the same procedure as at first, in each of the two intervals which it now divides, to have a minimised RMS departure in each.

This process of shortening the length of the curve spanned by a single piece continues automatically until a stipulated fit or close passage through every point is achieved. This result could be called 'faithful following' or 'tight fitting'. A relatively large permissible RMS departure leads to fewer pieces and a concatenated curve passing among rather than near and through the digitised points which are not control points.

The method can serve as either a *close fitting or a smoothing, slope-continuous* curve fitting technique according to the RMS departure set.

Treatment of vertical tangents is by rotation before processing followed by subsequent restoration. If the number of points in an interval is reduced to two by arc-shortening, a straight line is drawn between them. Fairly close digitisation is presumed.

As shown in the example below, the program prints out lists of coefficients of every piece found, together with the end limits of the interval which it spans.

Curves may be plotted. Except where selected and marked initially, no inflexions can appear (as they might with cubic expressions).

Refer to Chapter 5 of *Sculptured Surfaces in Engineering and Medicine* (Duncan and Mair, 1983) for more details.

Hardware:

IBM PC compatible personal computer running under DOS operating system with 640K of RAM and a hard disk or high density floppy disk.

Input:

The following prompts are requested by the user:

    Enter name of input file:
    Enter CTYPE:
    Coeffs: Enter 1 for x = f(y), 0 for y = f(x):
    Enter desired RMS:
    Enter 0 for 3pts & 2slopes; 1 for 5pts:

CTYPE is the curve type. If CTYPE is 'L', then a line is fitted through the points. Any other character will fit a conic through the points. This is important if one is using CURVFIT for the input of PROPDEV.

X and Y increments regulate the spacing of the interpolated points and the amount of shift by the program to adjust the mid-point to minimize the RMS error.

Coeffs determines if the user wishes the coefficients of the conic to be as a function of X or as a function of Y.

RMS is the maximum root mean square error allowed for the curve.

The input file format is as follows:
If there are no inflexion points, skip to *x1slope, y1slope*
    ninfl
infl1, infl2, ..., infln
    x1inf, y1inf
    x2inf, y2inf

    xninf, ynifn
    x1slope, y1slope
        x1, y1
        x2, y2

        xn, yn
    xnslope, ynslope

Explanation:

ninfl is the number of inflexion points on the curve
infl1, infl2, ..., infln are the numerical positions of inflexion points. They are determined by the position after the first *infl1* data points.
x1inf, y1inf is the first inflexion point.
x1slope, y1slope is the point to determine the starting slope of the curve.
xnslope, ynslope is the point to determine the ending slope of the curve.

Output:
The program will generate the following output files:
CURVFIT.COE contains the interval and the coefficients for each interval.
CURVFIT.PTS contains the interpolated points for the curves.

CURVFIT.PLT contains the plotting instructions for viewing the curve using the programs discussed in this book.

Example:

Enter name of the input file: *curve.dat*
Enter CTYPE: *P*
Enter X and Y increments: *0.1, 0.1*
Enter 1 for x = f(y); 0 for y = f(x): *0*
Enter desired RMS: *0.1*
Enter 0 for 3pts & 2slopes, 1 for 5pts: *0*

Note:

The user can use any of the viewing routines discussed in the book to plot the file CURVFIT.PLT

The user can also use CURVFIT.COE as part of the input for the program PROPDEV.FOR

Figure B2 Example of curve using CURVFIT.

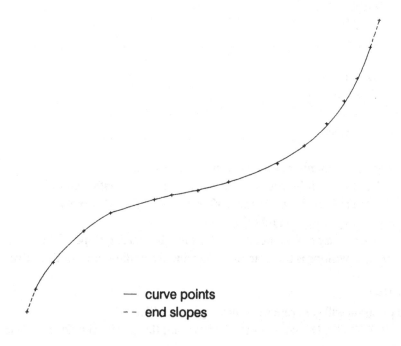

— curve points
- - end slopes

PROPDEV

Purpose:

The program PROPDEV develops parametric curves defining a surface to span a surface-patch bounded by four twisted space-curves. The space-curves and path are visualised as seated in Cartesian space and are represented in orthographic projection. The parameters used are fractions of the range of any two of the three coordinates along boundaries. These fractions are derived by proportional development, hence the name PROPDEV. After the first interpolation, the surface can be adjusted locally to produce a surface with some pre-determined non-boundary characteristics. This method of proportional development allows one patch of an irregular boundary surface to be interpolated. Connecting several patches will not have any slope continuity at the boundaries of the neighbouring patches.

Method:

The four bounding space-curves of a surface-patch as shown in the following figure may be assigned functional forms either in terms of a single parameter or by functions assigned in two planes of their orthographic projection. Alternatively, they may be determined by CURVFIT in two projections.

A closed continuous space-curve may be marked at four points to divide it into four boundary curves which are incidentally tangential to each other at those points.

A patch ABCD is shown in two projections (see Figure A2). Its boundaries are AB, CD, trending in the x direction and AC, BC, trending in the y and z directions. Each space-curve boundary is defined by two projections.

In the elevation we divide the x and z ranges of boundary curves in ratios from zero to one, α for z and β for x.

An example given below illustrates a case where the boundary curves have been arbitrarily made circular arcs and straight lines in their projected views. Often these curves will be derived from digitised sketches: then CURVFIT is used to derive a piecewise function for the boundaries and PROPDEV can make use of such functions for its execution.

Figure B3 Demonstration surface in plan and elevation views.

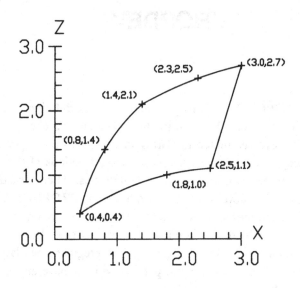

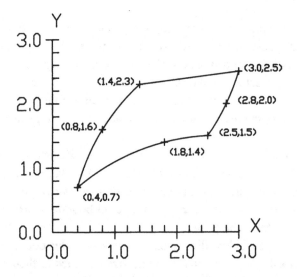

Hardware:

IBM PC compatible personal computer running under DOS operating system
with 640K of RAM and a hard disk or high density floppy disk.

Input:

The input is in the form of a data file whose format is as follows:

line 1: view description
line 2: number of ratios, zero if using previous ratio
line 3: curve name
line 4: list of ratios required
line 5: view name
line 6: curve type
line 7: starting and ending coordinates of the view
line 8: starting and ending coordinates of the curve
line 9: coefficient of the curve.

If the view has more than one curve (i.e. more than one curve to describe
the interval), then repeat lines 8 and 9. Then repeat lines 1 to 9 three times for
curves 2, 3, and 4 respectively.

The view description (line 1) consists of 8 characters which describe the
following boundary curves. The first two characters are either 'E_' for elevation, 'P_' for plan, or 'EE' for end elevation. The third character, which is
either 'X', 'Y' or 'Z', describes the independent variable and the fourth
character describes the dependent variable of the coefficients of the curve.
Therefore the first four characters describe the first two curves and the
second four characters describe the second two curves.

If the number of ratios (line 2) is zero, use the previous ratios and line 5
is omitted.

Curve name (line 3) is usually U0, U1, V0, V1. U0 corresponds to the
lower horizontal curve while U1 corresponds to the upper horizontal curve.
Similarly, V0 corresponds to the left vertical curve and V1 corresponds to the
right vertical curve.

The list of ratios (line 4) must start with zero and end with one.

The view name (line 5) can be 'E_' for elevation, 'P_' for plan, or 'EE' for
end elevation.

If the curve type (line 6) is a conic, a blank line suffices.

If the curve type is a straight line, a single character 'L' is placed at the first column in line 6. Line 7 contains the starting and terminating coordinates of the line in that particular view. Lines 8 and 9 are omitted.

The input data file format generates an initial surface. The modification of raising and/or lowering regions of the surface can be done in the form of the Mitchell solution of a circular plate deflection due to a load position somewhere on the plate. The parameters are the height or the load of the adjusted surface, the position of the load, and the angle of the load with respect to the yz plane.

Output:

The program generates two output files: surface point file and plot file. The surface point file lists each point per line by its x, y, z coordinates. The plot file shows various views of the surface as well as the final ratios after the adjustments. Various programs described here can be used to display the plot file.

Note:

The maximum number of ratios per direction is 64.

GEN7

Purpose:
Surface modelling using piecewise analytical compound surfaces.

Method:
A surface is defined by the intersection of a series of analytical surfaces over a topologically rectangular grid in the $x\,y$ coordinate plane. The analytical surfaces consist of ellipsoids (spheres can be thought of as ellipsoids with equal axes), elliptic paraboloids, hyperbolic paraboloids, cones, elliptic cylinders, tori, elliptical cylinders, and planes (see Figures B4 and B5). These analytical surfaces can be rotated and positioned anywhere on the user specified grid. The 'Method of Highest Point' is used to determine the z height of the surface at any given position in the $x\,y$ projection plane. See Chapter 9 of *Sculptured Surfaces in Engineering and Medicine* (Duncan and Mair, 1983) for a general discussion of this method.

Hardware:
IBM PC compatible personal computer running under DOS operating system with 640K of RAM and a hard disk or high density floppy disk.

Input:
FILE?
Enter the name of the data file to store the surface points.
ENTER FIELD DIMENSION X AND Y?
Enter the dimension of the working area in the plan view. The program assumes the lower left corner to be the origin. The number of points in each direction is one more than the field dimension divided by the increment.

The following prompts request the user to enter the number of surface pieces for a particular type of analytical piecewise surface (e.g. number of ellipsoids in the surface). If the number is 0, then the program requests the number for the next type of analytical piece (e.g. cone). If the number is greater than 0, then the program will request information of each piece such

125

Figure B4 Surface elements for the GEN7 program.

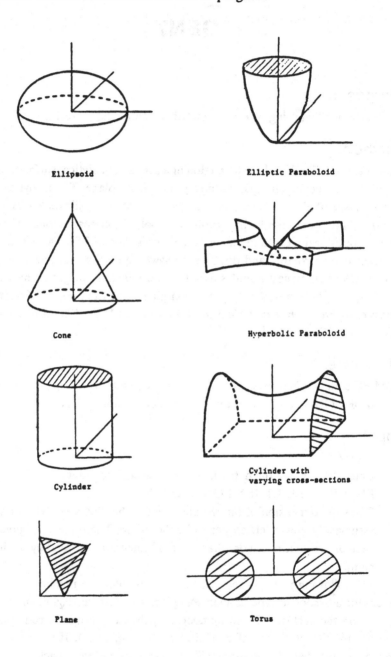

Figure B5 Input parameters for GEN7.

Surface Type	Characteristic Equation	User Inputs
Ellipsoid	$\dfrac{x^2}{a^2} + \dfrac{y^2}{b^2} + \dfrac{z^2}{c^2} = 1$ For spheres : $a = b = c = r$ (radius)	Centroid .... x_0, y_0, z_0 Semi-axes .... a, b, c Rotations .... $\theta_1, \theta_2, \theta_3$ Subdomain Limits .... x_{min}, y_{min} x_{max}, y_{max} Offlimit Height ..... z_{off} Truncation Height ... z_{tr}
Elliptic Paraboloid	$\dfrac{x^2}{a^2} + \dfrac{y^2}{b^2} = cz$	Vertex .... x_0, y_0, z_0 Semi-axes .... a, b, c Rotations .... $\theta_1, \theta_2, \theta_3$ Subdomain Limits .... x_{min}, y_{min} x_{max}, y_{max} Offlimit Height ..... z_{off} Truncation Height ... z_{tr}
Hyperbolic Paraboloid	$\dfrac{x^2}{a^2} - \dfrac{y^2}{b^2} = cz$	Vertex .... x_0, y_0, z_0 Major & Minor Axes .. a, b, c Rotations .... $\theta_1, \theta_2, \theta_3$ Subdomain Limits .... x_{min}, y_{min} x_{max}, y_{max} Offlimit Height ..... z_{off} Truncation Height ... z_{tr}
Quadratic Cone	$\dfrac{x^2}{a^2} + \dfrac{y^2}{b^2} + \dfrac{z^2}{c^2} = 0$ For circular Cones : $a = b = \tan\phi$ ϕ = semi-angle $c = 1$	Centroid .... x_0, y_0, z_0 Semi-axes .... a, b, c Rotations .... $\theta_1, \theta_2, \theta_3$ Subdomain Limits .... x_{min}, y_{min} x_{max}, y_{max} Offlimit Height ..... z_{off} Truncation Height ... z_{tr}
Quadratic Cylinder	$\dfrac{x^2}{a^2} + \dfrac{y^2}{b^2} = 1$ Length $= 2r_0$	Centroid .... x_0, y_0, z_0 Semi-axes .... a, b Half Length .... r_0 Rotations .... $\theta_1, \theta_2, \theta_3$ Subdomain Limits .... x_{min}, y_{min} x_{max}, y_{max} Offlimit Height z_{off} Truncation Height ... z_{tr}
Plane	$\dfrac{x}{a} + \dfrac{y}{b} + \dfrac{z}{c} = 1$	Intercepts .... a, b, c Subdomain Limits ... x_{min}, y_{min} x_{max}, y_{max} Truncation Height .. z_{tr}
Torus	$\left(\sqrt{x^2 + y^2} - a\right)^2 + z^2 = b^2$	Centroid .... x_0, y_0, z_0 Ring Radius .... a Tube Radius .... b Subdomain Limits .... x_{min}, y_{min} x_{max}, y_{max}
Tubular surface with parabolic profile	$z = \left(cx^2 + b\right)\sqrt{1 - \dfrac{y^2}{a^2}}$	Centroid .... x_0, y_0, z_0 Parameters .... a, b, c Rotation (z-axis) . θ_1 Subdomain Limits .... x_{min}, y_{min} x_{max}, y_{max} Offlimit Height ..... z_{off} Truncation Height ... z_{tr}

as the centre of the analytical piece, the major and minor axes, the limits within which this surface-piece is defined and the maximum height of the piece.

Output:

The output is a file containing the data points of the surface. The format is three real numbers per line which represent the x, y, z coordinates respectively.

Note:

The user can then use the output file as input to the program PLOT3D to display the surface with hidden line removal. See the PLOT3D documentation for usage.

Example:

This is an example to show the typical inputs requested by GEN7. The result of this particular set of inputs is shown in Figure B6.

```
file?demo.dat
ENTER FIELD DIMENSION  X AND Y? 3.4,2.8
ENTER INCREMENT D ? .1
NUMBER OF ELLIPSOIDS ( max 3 )? 1
   ENTER (X0,Y0,Z0) FOR ELLIP( 1 ) ............. ? 1,1.2,0
   ENTER A, B AND C, FOR ELLIP( 1 ) ............. ? .6,.8,1
   ENTER ROTATIONS 1, 2 AND 3, FOR ELLIP( 1 ) ... ? 0,0,0
   ENTER LOWLIMX, UPLIMX FOR ELLIP( 1 ) ... ? .4,1
   ENTER LOWLIMY, UPLIMY FOR ELLIP( 1 ) ... ? .4,1.2
   ENTER OFFLIMIT HEIGHT FOR ELLIP( 1 ) ... ? 0
   ENTER TRUNCATION HEIGHT FOR ELLIP( 1 ) . ? 99
Pausing .. Type 1 to alter input, any no. to continue ? 0
NUMBER OF EL. PARAB ( max 3 ) ? 0
NUMBER OF HYP. PARAB ( max 3 )? 0
NUMBER OF QUADRATIC CONE ( max 3 ) ? 1
   ENTER VERTEX (X0,Y0,Z0), FOR CONE( 1 ) .. ? 2,1.8,2
   ENTER A, B  AND C, FOR CONE( 1 ) ........ ? 1.2,1.2,2
   ENTER ROT 1, 2 AND 3, FOR CONE( 1 ) ..... ? 0,0,0
   ENTER TRUNCATION HT FOR CONE( 1 ) ....... ? 1.8
```

ENTER OFFLIMIT HT FOR CONE(1) ? 0
ENTER LOWLIMX AND UPLIMX FOR CONE(1) ? 1,3
ENTER LOWLIMY AND UPLIMY FOR CONE(1) ? 1.2,2.4
Pausing .. Type 1 to alter input, any no. to continue ? 0
NUMBER OF ELLIPTIC (CIRCULAR) CYLINDER (max 3) ? 2
ENTER CENTRE POINT (X0,Y0,Z0) FOR CYL(1) .. ? 2.2,1.2,0
ENTER A, B AND R0, FOR CYL(1) .... ? 1,.8,1.2
ENTER ROT 1, 2 AND 3 FOR CYL(1) ..... ? 0,90,0
ENTER LOWLIMX AND UPLIMX FOR CYL(1) .... ? 1,3.4
ENTER LOWLIMY AND UPLIMY FOR CYL(1) .... ? .4,1.2
ENTER OFFLIM HT FOR CYL(1) .... ? 0
ENTER TRUNCATION HT FOR CYL(1) .... ? 99
ENTER CENTRE POINT (X0,Y0,Z0) FOR CYL(2) .. ? 1,1.9,0
ENTER A, B AND R0, FOR CYL(2) .... ? .6,1,.7
ENTER ROT 1, 2 AND 3 FOR CYL(2) ..... ? 0,0,90
ENTER LOWLIMX AND UPLIMX FOR CYL(2) .... ? .4,1
ENTER LOWLIMY AND UPLIMY FOR CYL(2) .... ? 1.2,2.6
ENTER OFFLIM HT FOR CYL(2) .... ? 0
ENTER TRUNCATION HT FOR CYL(2) .... ? 99
Pausing .. Type 1 to alter input, any no. to continue ? 0
NUMBER OF PLANES ? 1
ENTER INTERCEPTS X,Y AND Z FOR PLANE(1) .. ? 100000,
100000,1
ENTER LOWLIMX AND UPLIMX FOR PLANE(1) .... ? 1,3.4
ENTER LOWLIMY AND UPLIMY FOR PLANE(1) .... ? 1.2,2.6
ENTER TRUNCATION HT FOR PLANE(1) .... ? 9
Pausing .. Type 1 to alter input, any no. to continue ? 0
NUMBER OF TORUS (max 3)? 0
NUMBER OF PARABOLIC ELLIPTICAL CYL (max 3)? 0
Program running ... starting time is 13:56:21
0 1 2 3 4 5 6 7 8 9 10 11 12 13 14 15 16 17 18 19 20 21
 22 23 24 25 26 27 28 29 30 31 32 33 34
ending time is 13:56:39

Figure B6 A result of the GEN7 program.

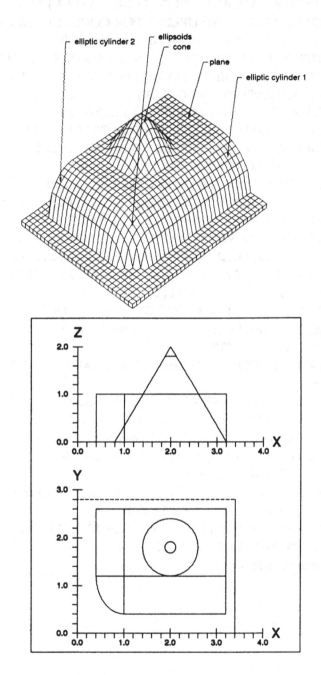

TRUEPERS

Purpose:

TRUEPERS is a general surface-fitting program to interpolate random three dimensional data points into a gridded surface of constant increments of x and y. The x increment need not necessarily be equal to the y increment. This program was written by J. Taylor, P. Richards, and R. Halstead of the Marine Sciences Division of Department of Energy, Mines and Resources, Canada. Dr. J. P. Duncan has obtained general release of this program. The program has been modified by K. K. Law to be executed on an IBM compatible personal computer.

Method:

Briefly, points x, y, z specified randomly by coordinates over an xy plane are used to interpolate or extrapolate z values at the nodes of a geometrically and topologically rectangular $x\,y$ grid. Local interpolation by iteration, using the equation of an elliptical paraboloid, and the fourth order differential equation governing the bending of plates of stiffness 'K', a freely selected quantity, are used to compute surface-point heights at nodes in the xy plane. The resulting surface may either be a smoothing one, passing smoothly among the points with overshoots, or may have the nature of a 'trampoline' with the given 'points' sticking up like rods to force z values at their plan positions. A smooth surface passes through these interpolated points (and the original random points).

Points so derived may be linked by straight lines to form orthogonal profiles which may be plotted perspectively from selected viewpoints by PLOT3D. Contours may be derived and plotted.

Hardware:

IBM PC compatible personal computer running under DOS operating system with 640K of RAM and a hard disk or high density floppy disk.

Input:

The following uppercase letters are displayed by the program.

131

ENTER INPUT FILE NAME:
File name containing the random points. The maximum number of random points is 2700. The format is one or more blanks followed by a real number followed by a comma and/or one or more blanks followed by a second real number and finally by a comma and/or one or more blanks and a third real number. The first real number represents the x coordinate, the second number the y coordinate, the third number the z coordinate. Each line should have only three real numbers to represent a point in space.

ENTER OUTPUT FILE NAME:
File name that will contain the interpolated gridded points. The format is the same as the input format.

ENTER NUMBER OF POINTS TO BE INTERPOLATED IN X:
An integer to indicate the number of points along the x direction. The maximum number of points in x is 127.

ENTER XMIN, XMAX:
The starting and ending value of the interpolated range.

ENTER NUMBER OF POINTS TO BE INTERPOLATED IN Y:
An integer to indicate the number of points along the y direction. The maximum number of points in y is 128.

ENTER YMIN, YMAX:
The starting and ending value of the interpolated range.

ENTER CAY, NRNG, NSM:
CAY is the stiffness of fit. The greater the CAY value, the tighter (smoother) the fit (real value expected). Normal CAY values range from 10 to 50.
NRNG is the number of adjacent points to consider during the interpolation process (integer value expected).
NSM is the number of smoothings the program will do (integer value expected).

ENTER INPUT ECHO (1 = YES):
If the user wishes to see the data being read, then type 1 . Otherwise type any integer.

BLANKING DATA? (1 = YES):
Blanking is to set all points outside a polygon to have a zero elevation. The polygon is defined in the blanking data file. The format of this file is one or more blanks followed by a real number followed by a comma or one or more blanks followed by a real number. The first real number corresponds to the x coordinate and the second number corresponds to the y coordinate respectively. The maximum number of blanking points is 420. If blanking is required, type 1 . Otherwise type any integer.

Output:
The output is stored in a file specified by the user. The format is the same as the input format.

Example:
Enter input file name: *truep.dat*
Enter output file name: *truep.tru*
Enter number of points to be interpolated in X: *11*
Enter Xmin, Xmax: *0.0, 5.0*
Enter number of points to be interpolated in Y: *11*
Enter Ymin, Ymax: *0.0, 5.0*
Enter CAY, NRNG, NSM: *20., 5, 4*
Enter input echo (1 = yes): *0*
Blanking data? (1 = yes): *0*

Notes:
The user may execute PLOT3D (see Figure B7) to view the surface.

If the surface has oscillations, the user may consider processing the data twice by TRUEPERS. For the first pass, the user specifies a low CAY and a coarse grid (perhaps twice the increment or half the number of points of the final grid) on the original data. For the second pass, the user inputs the result of the first pass as random data and specifies a high CAY and a finer grid. The second result will be smoother than that resulting from a single pass with a high CAY on fine grid data.

Figure B7 Example of TRUEPERS.

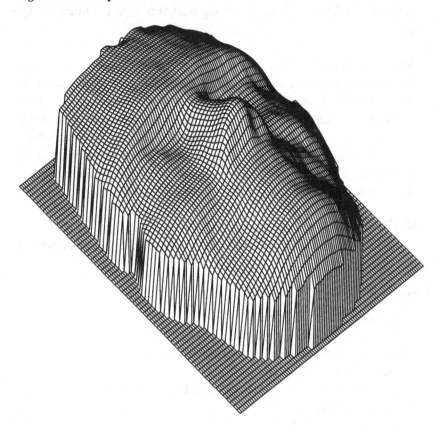

PLOT3D

Purpose:

PLOT3D is a program to plot a perspective view, with hidden line removal, of any surface represented by a grid of 3D points with an equal increment in the x and y direction such as the output of TRUEPERS. The equal increments of x do not need to be the same as y. This program was originally part of the program TRUEPERS, but was separated by K. K. Law. The output is a file of plotting instructions.

Method:

The program uses a ray tracing method to determine the visibility of each point.

Hardware:

IBM PC compatible personal computer running under DOS operating system with 640K of RAM and a hard disk or high density floppy disk.

Input:

The input data must be similar to the output of TRUEPERS (constant increment in x and y, but the increment of x need not be equal to y). The input format is the same as the output format of TRUEPERS. PLOT3D requests the following information:

ENTER THE NAME OF THE PLOT FILE:
The file to store the plotting instructions.

ENTER THE NUMBER OF POINTS IN X DIRECTION:
An integer which indicates the number of points in the X direction (max = 127).

ENTER THE NUMBER OF POINTS IN Y DIRECTION:
An integer which indicates the number of points in the Y direction (max = 128).

135

ENTER ZBASE, ZMAG, R, THETA, PHI:
ZBASE is the lowest level on the drawing
ZMAG is the scaling factor on the Z direction
R, THETA, PHI are the polar coordinates representing the eye position
for viewing the surface.

ENTER THE NTH POINT TO PLOT:
An integer to indicate the spacing of points to be plotted. Entering 1 will
plot every point. Entering 2 will plot every second point. And so on.

ENTER PAPER SIZE: XMIN, XMAX:
Two real numbers to indicate the range of the plotting area in the X
direction.

ENTER PAPER SIZE: YMIN, YMAX:
Two real numbers to indicate the range of the plotting area in the Y
direction.

ENTER NUMBER OF CONTOUR LEVELS (0 = NONE):
If no contours are desired, enter 0. If contours are desired, enter the
number of contours desired. Then PLOT3D will ask for the elevation of
each contour. The maximum number of contours is 10.

Output:

The output is a file of plotting instructions to control a plotter to plot the
perspective view. The format of a plot file is as follows:

ipen x-coord y-coord icolour
where *ipen* is an integer which indicates the pen status for the point.
ipen = 0 means pen down
ipen = 1 means pen up
x-coord, y-coord are the X and Y coordinates of the point to be plotted.
icolour is an integer to indicate the pen number.

Example:

Enter name of input file: *truep.tru*
Enter name of plot file: *truep.plt*
Enter the number of points in X direction: *11*

Enter the number of points in Y direction: *11*
Enter ZBASE,ZMAG,R,THETA,PHI: *-.1,1.,30.,45.,45.*
Enter every Nth point to plot: *1*
Enter paper size: Xmin, Xmax: *0., 10.*
Enter paper size: Ymin, Ymax: *0., 10.*
Enter number of contour levels (0 = none): *0*

Note:

The plot file is compatible with any of the displaying programs described in this book. Other graphic screens and plotters must follow the above plotting commands and write a driver for each device.

PLOT3D can be used as input from the output files of the following programs:
- GEN7
- CURVATUR
- TRUEPERS
- data file of errors from SUPERSUE

TERMPLOT is a driver for an IBM Colour Card, Hercules Monochrome, and IBM Enhanced Colour display. HPPLOT is a driver for the HP plotter using the HP-GL commands. HIDMP is a simple driver for the Houston Instrument DMP 50 series plotter. With any other plotter, the user must write a driver for his or her plotter.

PROJ

Purpose:

Program PROJ projects an object defined in *xyz* space onto the vertical plane. The points may be plotted by connecting every point with straight lines or may be plotted as a point. No hidden lines are removed.

Method:

The equations for the position of the object in the picture plane can be derived using simple geometry. The relationships between picture plane position (x_s, z_s) and a point on the object given by (x_p, y_p, z_p) to the user's eye position given by (x_e, y_e, z_e) are:

$$x_s = x_p + y_p (x_e - x_p) / (y_p - y_e)$$
$$z_s = z_p + y_p (z_e - z_p) / (y_p - y_e)$$

The derivation of the above relationships can be found in Chapter 3 of *Sculptured Surfaces in Engineering and Medicine* (Duncan and Mair, 1983).

Hardware:

IBM PC compatible personal computer running under DOS operating system with 640K of RAM and a hard disk or high density floppy disk.

Input:

The program prompts the user for the following data:

Enter name of data file:
Enter the name of the data file in which the object coordinates are stored.

Enter name of plot file:
Enter the name of the plot file for storing the plotting instructions.

Enter eye position:
The user's eye position for viewing the object. The eye position should have a negative value in the *y* direction.

Do you want to plot points? (1 = yes)
If the user wishes to see the points without lines connecting them, then enter 1. Otherwise enter any integer.

Data file format:

Each line in the data file contains three real numbers separated by one or more blanks. The first represents the *X*-value, second the *Y*-value, third the *Z*-value.

Output:

Output consists of a plot file (see the output of the PLOT3D program for the format).

Note:

The user may use TERMPLOT, HPPLOT, HIDMP, or any other plotting routine the user may have as long as the input follows the format described in the output section of PLOT3D.

TERMPLOT

Purpose:
Displays on a graphic screen any plot file produced by the programs described in this book. The format is the same as described in the documentation of PLOT3D.

Method:
Simple mapping of the plot file coordinates to the device coordinate system.

Hardware:
IBM PC compatible personal computer running under DOS operating system with 640K of RAM and a hard disk or high density floppy disk and an IBM compatible Colour Graphic Adapter (CGA), IBM compatible Enhanced Graphic Adapter (EGA), or Hercules Monochrome Graphics.

Input:
To execute TERMPLOT, type
    TERMPLOT plotfile
    where *plotfile* is the name of the plot file.

The program will prompt the user for the following infomation:
ENTER XMIN, XMAX:
Enter the window limits in *X. Xmin* and *Xmax* must be separated by one or more blanks.

ENTER YMIN, YMAX:
Enter the window limits in *Y. Ymin* and *Ymax* must be separated by one or more blanks.

Output:
Picture of the plot file.

Example:

To run TERMPLOT:
> *TERMPLOT example.plt*
> Enter Xmin, Xmax: *0.0 5.0*
> Enter Ymin, Ymax: *1.0 6.7*

Notes:

If the user does not have a plotter and wishes to have a hardcopy, he or she may run GRAPHICS on a DOS disk before running TERMPLOT. After TERMPLOT displays the output, the user can press <Shift> <PrtSc> to obtain a screen dump onto a dot matrix printer.

HPPLOT

Purpose:

A program to read the plotting instruction from the format described in the output section of PLOT3D and drives any plotter which supports the Hewlett Packard Graphic Language (HP-GL). The plotter must be connected to the COM1: . All the graphical outputs described in the POLYHEDRAL NC® use this format.

Hardware:

IBM PC compatible personal computer running under DOS operating system with 640K of RAM, a hard disk or high density floppy disk, a RS-232 asynchronous communication port and a plotter that supports Hewlett Packard Graphic Language (HP-GL) instructions.

Input:

Before executing HPPLOT, users must set up the correct parameters on their communications port. This includes using the MODE command in DOS to initialise the asynchronous port.

To execute HPPLOT, type
HPPLOT
The program will prompt the user as follows:

Enter name of the plotfile:
Plotfile is the name of the plotting instructions.

ENTER XMIN, XMAX, YMIN, YMAX:
Enter the window limits in *X* and *Y*. *Xmin, Xmax, Ymin* and *Ymax* must be separated by one or more blanks. HP-GL only accepts integer values. If real numbers are used unpredictable results may occur.

Enter pen velocity:
Pen velocity can range between 0.1 to 38.1cm/s.

Enter title:

Title is a sequence of characters that the user types and which will be written on the drawing. A maximum of 40 characters is permitted. A space in the title will terminate the title.

Output:

Plotter moves and draws the picture.

Example:

To run HPPLOT:

MODE COM1:9600,N,8,1,P

HPPLOT

Enter Xmin, Xmax, Ymin, Ymax: *0.0 5.0 1.0 6.7*

Enter pen velocity: *30*

Enter title: *test_demo*

HIDMP

Purpose:

A program to read the plotting instruction from the format described in the output section of PLOT3D which will drive any plotter which supports the Houston Instrument DMP 40 series plotters. The plotter must be connected to the COM1:. All the graphical outputs described in the POLYHEDRAL NC® use this format.

Hardware:

IBM PC compatible personal computer running under DOS operating system with 640K of RAM, a hard disk or high density floppy disk, a RS-232 asynchronous communication port and a plotter that supports the Houston Instrument DMP 40 series plotter instructions.

Input:

Before executing HIDMP, the user must set up the correct parameters on the communications port. This includes using the MODE command in DOS to initialise the serial port. Refer to the Plotter Owner's Manual.

To execute HIDMP, type
HIDMP
The program will prompt the user for the following information:

Enter name of the plotfile:
plotfile is the name of the plotting instructions.

ENTER XMIN, XMAX, YMIN, YMAX of plot file:
Enter the window limits in X and Y. *Xmin, Xmax* (*Ymin* and *Ymax* must be separated by one or more blanks).

ENTER XMIN, XMAX, YMIN, YMAX of plotter unit:
Enter the viewport limits in X and Y. *Xmin, Xmax* (*Ymin* and *Ymax* must be separated by one or more blanks).

144

Output:

Plotter moves and draws the picture.

Example:

To run HIDMP:

HIDMP

Enter Xmin, Xmax, Ymin, Ymax of plotfile: *0.0 5.0 0.0 8.0*

Enter Xmin, Xmax, Ymin, Ymax of plotter unit: *10 10 810 1010*

CURVATUR

Purpose:

The program assumes that coordinated points in a geometrically and topologically rectangular array are points of a continuous analytical surface. The modulus of the maximum principal curvature at each node is calculated by *numerical calculus* of the standard Gaussian expressions for such curvature in terms of the *mean and Gaussian curvatures*. Perspective plots of these values drawn via PLOT3D indicate local regions where positive curvatures occur and, by comparison of the related radius of curvature with any tool being used, where interference may occur and where tool retraction will be applied by either NEWERSUE or SUPERSUE. This information also assists the choice of suitable tool diameters for the surface.

Method:

Using the coordinated surface-point data produced by TRUEPERS or data obtained by modern optical measuring systems arranged in a similar array, the maximum principal curvature at nodes of a smooth surface represented approximately by the data may be found by using finite difference (numerical) calculus.

The reciprocal of the greatest positive value of principal curvature K_1 gives the smallest local radius of curvature. Spherically ended cutting tools of any greater radius than this, if caused to touch a local facet, would interfere with neighbouring facets unless withdrawn as done by NEWERSUE. So if it is desired to cut a complete surface patch with a single tool in one pass, the curvature computation indicates the maximum tool size that can be used.

If, for some reason, the largest tool to be used has been arbitrarily determined, the maximum curvature it can handle is found as the reciprocal of its radius. Any surface-curvature larger than this marks out a region of the surface where there is potential for interference; where withdrawal of the larger tool of a 'cascade' is necessary and where smaller tools must visit later to cut the surface more accurately.

Perspective plots of local maximum principal curvature of a numerically defined surface can be plotted through PLOT3D, taking curvature instead of the coordinate z as the ordinate. Also, any curvature greater than a nominated value may be plotted to reveal graphically where the potentially interfering regions of a surface are.

The program CURVATUR is useful in providing guiding information in the use of both NEWERSUE and SUPERSUE; it has thus been kept as an independent program.

Input:

The program prompts the user for the following information:

Enter name of data file:
Data file is any file whose format is the same as the output of TRUEPERS.

Enter name of output curvature file:
This file contains the curvature at each point on the surface. This file can be used as input to PLOT3D to view the nature of the curvature of the surface.

Enter minimum curvature to view:
Minimum curvature is the reciprocal of the tool radius being used. For example, if the user is using a half inch tool, then the minimum curvature is 2 (reciprocal of one-half). At any point, curvature greater than 2 will be displayed and curvature less than 2 (including less than 0) will be displayed as zero.

Output:

The output file shows the *xy* position and the curvature at that position. The format is the same as the output format of TRUEPERS. The output file can be used for input to the PLOT3D program to remove the hidden lines and view the curvature at any point on the surface.

NEWERSUE

Purpose:

NEWERSUE computes the Cutter Location Data (CLD) for a spherically ended tool which is caused to traverse and machine by pointing the whole domain of a surface specified by stored data for the vertices of an approximating polyhedron arranged above a topologically rectangular array in the xy coordinate plane. NEWERSUE comes in two forms: one for small arrays (45 by 45) with program data stored in memory and another for larger arrays with program data stored on disk. NEWERSUE should be used if the distance between data points is near the size of the tool radius.

Method:

The triangular facets of the approximating polyhedron in this array are visited systematically and the tool is retracted if necessary in the positive z direction to avoid its cutting (gouging) any facet within reach when it is addressed to a principal facet at a point in its progress over the field.

The program computes all the data necessary for a tool of smaller diameter to visit subsequently all facets missed by preceding larger tools. This is automatically computed for a 'cascade' of tools, each half the diameter of the preceding one. By this means, interference is avoided by ensuring that the tool does not 'gouge' any plane facet while a surface of variable curvature is correctly machined. (This is in contrast to SUPERSUE which causes the tool to avoid *points* rather than *planes*.)

NEWERSUE has the option to generate a file of errors (difference between the actual surface and the surface machined by a given tool of specified diameter) due to interference. The errors can be plotted using the PLOT3D. program

NEWERSUE is the latest version of its predecessors, SUMAIR and NEWSUE. Each version contains the features of its previous version. The detailed general theory of these programs is given in the *Sculptured Surfaces in Engineering and Medicine* (Duncan and Mair, 1983).

Hardware:

IBM PC compatible personal computer running under DOS operating system with 640K of RAM and a hard disk or high density floppy disk.

Input:

Enter input file name:

input file is the data file the user wishes to machine

Enter output file name:

output file is the file which stores the Cutter Location Data (CLD)

Enter origin:

origin is the distance which the tool is set relative to the data origin

Enter input scaling factor FX, FY, FZ:

FX, FY, FZ are the input data scaling factors

Enter RAD:

RAD is the radius of the largest tool being used. The program may select a smaller tool on its later pass

Enter output scaling OFX, OFY, OFZ:

OFX, OFY, OFZ are the output data scaling factors

Check interference (1 = not check, 0 = check):

Enter number of pass (1 or 2):

Enter calculation tolerance (0 means 0.001):

Enter smallest tool used (smallest = 0.03125):

Enter print switch (0 = off, 1 = on):

this switch will print information about interference

Enter data output: (1 = tool position, 2 = centroids):

Enter data size in X and Y:

Enter part of X to machine, (IM1,IM2) (IM1 = 0 means all):

Enter part of Y to machine, (JM1,JM2) (JM1 = 0 means all):

Error analysis (1 = yes, 0 = no):

IGRID is a switch to indicate if the data are a regular grid (that is increment in X is constant, and increment in Y is constant) IGRID = 0 if regular grid and IGRID = 1 if not

ID is a switch to indicate how the diagonal of the triangle of each grid is divided. If ID = 0 or 1 then the diagonal of the grid is divided in this direction: / ; if ID = 2 then the diagonal is in this direction: \ . Note: if NOP = 2 then ID should be 0

Output:

The output file format is as follows:

```
npt1   rad1   mode
x11  y11   z11
x12  y12   z12
 |    |    |
x1n  y1n   z1n
npt2   rad2
x21  y21   z21
x22  y22   z22
 |    |    |
x2n  y2n   z2n
and so on
```

where

npt1 is the number of points for the first tool

rad1 is the radius of the first tool to use

mode is the method to visit each point

if mode = 0, visit each point directly

if mode = 1, visit a point and then raise tool to non-cutting position before going to next point

npt2 is the number of points for the second tool

rad2 is the radius of the second tool to use

The error data file has the same format as the output of TRUEPERS except the Z-value is replaced by the error at the position X, Y. If the data are a topologically and geometrically rectangular, the error data file can be used as input for the PLOT3D program to view the nature of material left behind by a tool of a given radius.

SUPERSUE

Purpose:

SUPERSUE is a machining program based on the passage of a spherically-ended tool over a geometrically rectangular grid, the tool axis being located at a node of such a grid at each 'pointing' of the surface being cut. The tool height, z, is adjusted so that no spatial surface-node is ever within the volume of the tool-sphere. Thus it has the character of point avoidance rather than plane avoidance. Computations for interference are rather simpler than those used in NEWERSUE. After the first pass, SUPERSUE reduces the tool diameter by one-half. The program calculates the new tool position for those points which were not able to be machined in the previous pass. This re-machining allows the milling machine to use a large tool to remove the bulk of the material and use a smaller tool to machine the finer details of the model. SUPERSUE continues to reduce the tool diameter by one-half until the specified accuracy is achieved or until the tool is smaller than the mininum tool diameter specified by the user. In which case, the minimum tool diameter is used and the program terminates. SUPERSUE should be used when the distance between data points is much smaller than the tool radius.

Method:

SUPERSUE visits all the nodes of the geometrically rectangular grid. At every node, the program will check all neighbouring nodes under the projected area of the tool shank. If any of the neighbouring nodes were inside the *volume* of the tool, interference would occur. Then the tool is lifted sufficiently such that interference is avoided. The process is repeated for all neighbouring nodes within reach. In this way, the program monitors each node and records the smallest difference in height between what was machined and the original surface. This difference is known as the error of the surface for a given tool. If a smaller tool is used, less error results; however, more machining time is necessary for a given surface. The errors are stored in a file for later plotting by PLOT3D if necessary or for comparison by NEWERSUE.

Hardware:

IBM PC compatible personal computer running under DOS operating system with 640K of RAM and a hard disk or high density floppy disk.

Input:

The program prompts the user for the following information:

Enter input file name:
Data file contains the node points of the surface. The format is identical to the output format of TRUEPERS.

Enter name of tool position file:
Tool file is the file to store the cutter location data (CLD). The output format is identical to the output format of TRUEPERS.

Enter NX:
NX is the number of lines in the data file (max = 127).

Enter NY:
NY is the number of points per line in the data file (max = 128).

Enter every Nth point to machine in X:
If Nth point is 2, program will machine every other X point.

Enter every Nth point to machine in Y:
If Nth point is 2, program will machine every other Y point.

Enter tool diameter:
The diameter of the tool the user is using.

Enter machining tolerance:
The tolerance is the smallest unit of movement of the milling machine.

Enter minimum tool diameter:
Minimum tool diameter is the smallest tool that the user is using.

Enter error file name:

Error file contains the difference in height of what is machined and what the actual surface is. This file can be used as input to PLOT3D to view the nature of error for a given tool.

Output:

The output is a file of tool positions in a format similar to the output format of NEWERSUE. The first line is a header line which contains three numbers. They correspond to the number of points in this set, the tool radius for this set and the mode to machine for this set respectively. If the mode is 0, machining is done by visiting every point directly. If the mode is 1, machining is done by touching a point and then raising the tool to a non-cutting position before proceeding to the next point. Another header line may appear after the first set of points if re-machining is required.

If error analysis is done, a file of points is generated. The format is similar to the output format of TRUEPERS except the z-value is replaced by the error at each x, y position.

Note:

The file containing the errors can be plotted with hidden line removal using the PLOT3D program as input.

GCODE

Purpose:

GCODE is used to convert the Cutter Location Data (CLD) to a code which the controller requires for instructing a numerically controlled (NC) milling machine. Different controllers use different formats for controlling. GCODE follows the EIA RS-274D standard and can be used by many FANUC controllers. This standard consists of G-codes for general motion control, M-codes for miscellaneous functions, and various other codes. If the user's NC machine does not follow the EIA RS-274D standard, then another program must be written to comply with the other standard.

Hardware:

IBM PC compatible personal computer running under DOS operating system with 640K of RAM and a hard disk or high density floppy disk.

Input:

The program prompts the user for the following information:

Enter units? (1 = inch, 0 = mm)
If the user is using metric units, then enter 0, otherwise enter 1

Enter feedrate? (real)
Feedrate must be in the same units chosen above

Enter spindle speed? (int)
Spindle speed in revolution per min (RPM)

Enter direction? (1 = cw, 2 = ccw)
If the user wishes the spindle to spin clockwise, then enter 1, otherwise enter 2

Enter NC program number?
An integer the user wishes to call the NC program

Enter starting tool position from origin?
The coordinate of the tool relative to the model origin. This is used to set the G92 code

Enter size of material?
Enter size starting with X, Y, Z

Enter name of input file:
Data file to convert to the EIA RS-274D standard

Enter name of output file:
Output file to store the NC commands for transferring to NC machine

Note:

The communications link between the computer and the NC machine is not part of the POLYHEDRAL NC package because different controllers may use different communications protocol.

VCAM

Purpose:

VCAM is a program to calculate the volume, centroid, area and the first and second moments of inertia of a given solid with homogeneous density. The solid may be concave or convex (refer to Chapter 6 for details).

Method:

The solid is represented by arbitrary lines of latitude and longitude analogous to those of the planet Earth. There must be the same number of points in each latitude. The surface area is determined by the summation of all the triangular facets of the approximating polyhedron. The volume is determined by the summation of all the tetrahedra from the three vertices of a triangular facet and from the fourth vertex of the mass centroid of the polyhedron.

Hardware:

IBM PC compatible personal computer running under DOS operating system with 640K of RAM and a hard disk or high density floppy disk.

Input:

The program prompts the user for the following information:

Please type in the data file name:
data file contains the points of the solid. The file should contain three real numbers per line which corresponds to the x, y, z coordinates respectively. The points are ordered such that the first N points are of a given latitude. The next N points are of the next latitude and so on. The should be M numbers of set of latitudes

Enter M (longitude), N (latitude):
M is the number of latitude (max = 127).
N is the number of longitude (max = 127).

156

Output:

The following information is given:

- the filename of the input file
- the number of latitudes and number of points per latitude
- the volume of the solid
- the surface area
- the mass centroid
- the moment of inertia in the given orientation
- the product of inertia in the given orientation
- the principal moment of inertia
- the axes orientation for the principal moment of inertia

REFERENCES

Boulanger, P., Evans, K. B., Rioux, R. and Ruhlmann, L. (1986) 'Interface between a 3D laser scanner and a CAD/CAM system', *Soc. Manuf. Engs., 5th Can. CAD/CAM & Robotic Conf.*, Toronto, June, 1986, paper MS86-731, pp. 1-7.

Dev, P., Wood, S., White, D. N., Young, S. W. and Duncan, J. P. (1983) 'An interactive graphics system for planning reconstructive surgery', *Proc. Nat. Comp. Graphics Assoc. Conf.*, Chicago, pp. 130-135.

Duncan, J. P., Lau, Y. K. and Steeves, A. O. (1984) 'Designing and machining violin top plates', *Proc. Int. IASTED Conf.*, Nice, France, June 19-21, 1984, pp. 38-42.

Duncan, J. P., Law, K. K. and Steeves, A. O. (1985) 'Computer-aided machining of terraced models', *Proc. 10th Can. Cong. Applied Mechanics* (CANCAM 85), London, Ontario, pp. E9-E10.

Duncan, J. P. and Mair, S. G. (1977) 'Anti-interference features of polyhedral machining', Proc. PROLAMAT 76, book, *Advances In Computer-aided Manufacture*, D. McPherson, ed., North-Holland, New York, pp. 181-195.

Duncan, J. P. and Mair, S. G. (1983) *Sculptured Surfaces In Engineering And Medicine*, Cambridge University Press, Cambridge.

Duncan, J. P., Wild, P. M. and Hoemberg, P. M. (1984) 'Automatic metrology and machining of arbitrary surfaces', Soc. Manuf. Engs., Paper MS84-945, *Conf. on Sensor Technology for Untended Manufacture*, Chicago, USA.

Duncan, J. P., Zhang, Z. T. and Steeves, A. O. (1986) 'Automatic testing for collision in robotics', *Proc. Int. Conf. of Computer-aided Prod. Eng.*, University of Edinburgh, A/1 2-4, 1986, pp. 301-304.

Lockwood, E. H. (1961) *A Book Of Curves*, Cambridge University Press, Cambridge.

Trimmer, H. G. and Stern, J. M. (1980) 'Computation of global geometric properties of solid objects', *Computer Aided Design*, 12(6), pp. 301-303.

INDEX

Printed in the United States
By Bookmasters